Shallow Lakes

Developments in Hydrobiology 3

DR. W. JUNK BV PUBLISHERS   THE HAGUE – BOSTON – LONDON   1980

Shallow Lakes
Contributions to their Limnology

Proceedings of a Symposium, held at Illmitz (Austria),
September 23–30, 1979

Edited by
M. DOKULIL, H. METZ AND D. JEWSON

DR. W. JUNK BV PUBLISHERS THE HAGUE – BOSTON – LONDON 1980

Distributors

for the United States and Canada

Kluwer Boston, Inc.
190 Old Derby Street
Hingham, MA 02043
U.S.A.

for all other countries

Kluwer Academic Publishers Group
Distribution Center
P.O. Box 322
3300 AH Dordrecht
The Netherlands

Library of Congress Cataloging in Publication Data   CIP

Shallow Lakes. Contributions to their Limnology.
  (Developments in Hydrobiology, v. 3)
  includes Index.
  1. Limnology – congresses.
  I. Dokulil, M.   II. Metz, H.   III. Jewson, D.   IV. Series.
  QH96.A3S53   551.48'2   80-15017

ISBN-13: 978-94-009-9208-5   e-ISBN-13: 978-94-009-9206-1
DOI: 10.1007/978-94-009-9206-1

Cover design: Max Velthuijs

FOREWORD

The Symposium on Shallow Lakes, held from 23rd–30th, September 1979, at the Biological Research Station, Illmitz (Austria), was intended to give an insight into current European research on shallow lakes. The reason for the restriction to European participants was firstly to gather as much information as possible on investigations in one geographic area, and secondly the limited time and space available.

Since shallow lakes pose a number of problems specifically related to their depth, several symposia have been devoted to this subject. Meetings like the Symposium on the Limnology of Shallow Waters in Tihany (Hungary), in 1973 and the Symposium 'Flachseeforschung' in Steinhude (Fed. Rep. of Germany), in 1974 stressed the need for further communication amongst limnologists working in this field. Moreover several international projects, like the OECD-Eutrophication-Program and the MaB-Project, have included certain aspects of shallow lake limnology. It is hoped by the editors that the proceedings presented here will stimulate further research and a greater exchange of information in this field.

Our special gratitude is due to the provincial government of Burgenland for their financial support of the Symposium, as well as to those supporting the meeting in various ways. Moreover we wish to thank the directors of the two organizing institutions, Hofrat Dr. F. Sauerzopf (Biological Research Station Illmitz) and Prof. H. Löffler (Institute for Limnology of the Austrian Academy of Sciences). We wish to extend our sincere thanks to all participants, contributors, and to all our colleagues for their essential help in many ways. Dr. L. Hammer, Dr. B. Schuster, I. Gradl, and E. Köllner deserve special mentioning for their assistance.

The Editors

Biological Research Station, Illmitz.

CONTENTS

3. Lake Balaton, Hungary

4. Miscellaneous

4.1 Chemical aspects

4.2 Botanical aspects

4.3 Bacteriological aspects

4.4 Zoological aspects

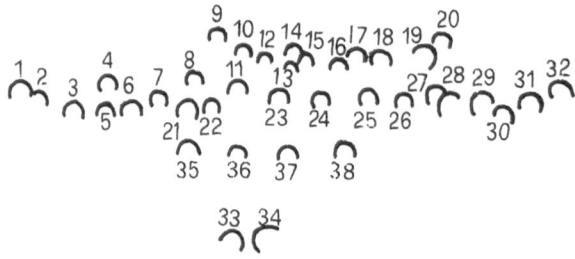

1: F. Sauerzopf; 2: B. Thaler; 3: L. Hammer; 4: B. Entz; 5: L. Stella; 6: I. Karpati; 7: R. Schröder; 8: R. Youngman; 9: V. Ilmavirta; 10: S. Dudzinski; 11: D. Reinhardt; 12: A. Herzig; 13: F. Wawrik; 14: H. Toivonen; 15: W. Donderski; 16: C. Abella; 17: E. Schulze; 18: D. Ernst; 19: H. Fleckseder; 20: K. Schwarz; 21: G. Dietz; 22: H. Schröder; 23: B. Georgi; 24: G. Deisinger; 25: H. Metz; 26: J. Šapkarev; 27: D. Jewson, 28: R. Guerrero; 29: B. Riemann; 30: L. Schultz; 31: N. Schulz,; 32: M. Dokulil; 33: J. Poltz; 34: J. Merkt; 35: G. Petrović; 36: M. Wallsten; 37: E. Stella; 38: R. Martínez.

LIST OF PARTICIPANTS

ABELLA, C. – Autonomous University of Barcelona, Bellaterra, Barcelona, Spain.

DEISINGER, G. – Kärtner Institut für Seenforschung, Flatschacherstraße G. Dietz, A-7142 Illmitz, Austria.

DOKULIL, M. – Institut für Limnologie der Österreichischen Akademie der Wissenschaften, Berggasse 18, A-1090 Wien, Austria.

DONDERSKI, W. – Laboratory of microbiology, Institute of Biology, Nicolaus Copernicus University, Gagarina 9, PL-100 Toruń, Poland.

ENTZ, B. – Biological Research Institute of the Hungarian Academy of Sciences, H-8237 Tihany, Hungary.

ERNST, D. – Universität Hannover, Institut für Biophysik, Herrenhäuserstraße 2, D-3000 Hannover 21, F.R.G.

FLECKSEDER, H. – Institut fïr Wasserversorgung, Technische Universität Wien, Karlsplatz 13, A-1040 Wien, Austria.

GEORGI, B. – Universität Hannover, Institut für Biophysik, Herrenhäuserstraße 2, D-3000 Hannover 21, F.R.G.

GUERRERO, R. – Autonomous University of Barcelona, Bellaterra, Barcelona, Spain.

HAMMER, L. – Biologisches Forschungsinstitut für das Burgenland, Biologische Station Illmitz, A-7142 Illmitz, Austria.

HERZIG, A. – Institut für Limnologie der Österreichischen Akademie der Wissenschaften, Berggasse 18, A-1090 Wien, Austria.

ILMAVIRTA, V. – University Helsinki, Dept. Botany, Unioninkatu 44, SF-00170 Helsinki 17, Finland.

JEWSON, D. – The New University of Ulster, Limnology Laboratory, Traad Point, Drummenagh, Magherafelt, Londonderry, Northern Ireland.

KARPATI, I. and V. – Agrarwissenschaftliche Universität, Lehrstuhl für Botanik und Pflanzenphysiologie, Deák F.U. 16, H-8361 Keszthely, Hungary.

LÖFFLER, H. – Limnologisches Institut der Österreichischen Akademie der Wissenschaften, Berggasse 18, A-1090 Wien, Austria.

MARTÍNEZ, R. – University Granada, Dept. Ecology, Granada, Spain.

MONTESINOS, E. – Autonomous University of Barcelona, Bellaterra, Barcelona, Spain.

METZ, H. – Biologisches Forschungsinstitut für das Burgenland, Biologische Station Illmitz, A-7142 Illmitz, Austria.

MERKT, J. – Niedersächsiches Landesamt für Bodenforschung, Box 51 0 153, D-3000 Hannover 51, F.R.G.

NEUHUBER, F. – Institut für Limnologie der Österreichischen Akademie der Wissenschaften, Berggasse 18, A-1090 Wien, Austria.

Neuwirth, F. – Zentralanstalt für Meteorologie, Hohe Warte 38, A-1190 Wien, Austria.

Olah, J. – Fisheries Research Institute, H-5541 Szarvas, Hungary.

Ostendorp, W. – Limnologisches Institut, Konstanz-Egg, Mainaustr. 212, D-7750 Konstanz, F.R.G.

Petrovic, G. – Institut für biologische Untersuchungen, Str. 29 November 142, YU-11 000 Beograd, Yugoslavia.

Poltz, J. – Limnologische Untersuchungsstelle Steinhude, Achternümme 6, D-3050 Wunstorf 2, F.R.G.

Reinhardt, D. – Universität Hannover, Institut für Strahlenbotanik, Herrenhäuserstraße 2, D-3000 Hannover, 21, F.R.G.

Riemann, B. – Freshwater Biological Laboratory, Univ. Copenhagen, 51 Helsingørsgade, DK-3400 Hilleröd, Denmark.

Ripl, W. – Institute of Limnology, University Lund, S-22003 Lund 3, Sweden.

Sauerzopf, F. – Biologisches Forschungsinstitut für das Burgenland, Biologische Station Illmitz, A-7142 Illmitz, Austria.

Šapkarev, J. A. – Institut of Zoology, Biological Faculty, University of Skopje, YU-91000 Skopje, Yugoslavia.

Schröder, H. and R. – Landesanstalt für Umweltschultz, Baden-Würtemberg, Institut für Seenforschung und Fischereiwesen, D-7752 Insel Reichenau, F.R.G.

Schulz, Z. and N. – Kärntner Institut für Seenforschung, Flatschacherstraße 70, A-9020 Klagenfurt, Austria.

Schulze, E. – Universität Hannover, Institut für Strahlenbotanik, Herrenhäuserstraße 2, D-3000 Hannover 21, F.R.G.

Schwarz, K. – Bundesinstitut für Gewässerforschung und Fischererreiwirtschaft, A-5310 Mondsee, Scharfling, Austria.

Stella, E. – Universita di Roma, Istituto Zoologica, Viale dell' Universitá 32, I-00100 Roma, Italy.

Tátrai, I. – Biological Research Institute of the Hungarian Academy of Sciences, H-8237 Tihany, Hungary.

Thaler, B. – Biologisches Landeslabor der Provinz Bozen, I-39 100 Bozen, Italy.

Toivonen, H. – University Helsinki, Dept. Botany, Unioninkatu 44, SF-00170 Helsinki 17, Finland.

Vörös, L. – Biological Research Institute of the Hungarian Academy of Sciences, H-8237 Tihany, Hungary.

Wallsten, M. – Institute of Limnology, Box 557, S-75122 Uppsala, Sweden.

Wawrik, F. – Biologisches Labor Waldviertel, A-3943 Schrems/Gebharts, Austria.

Weiss, R. – Österreichischer Wasserwirtschaftsverband, An der Hülben 4, A-1010 Wien, Austria.

Youngman, R. – Water Research Centre (Medmenham Laboratory), P.O.B. 16 Medmenham, Marlow, Bucks SL7 2HD, Great Britain.

1. STEINHUDER MEER, FED. REP. OF GERMANY

BACKGROUND DATA

Latitude: 52°28'N
Longitude: 9°20'E
Altitude: NN + 37.80 m
l (km) ca 8
b (km) ca 4.5
L (km)
L_D
Name of the main tributary (groundwater: ca 60% of total inflow)
Average inflow m³/sec.
Average outflow m³/sec. ca 0.5
Theoretical retention time ca 2.5 years

origin: glacial: thermokarst (+ deflation ?)
z (m) 2.80
$\bar{z}$ (m) 1.35
$\bar{z}/z \sim 1:2$
V (km³) 0.04
A (km²) 29.1
A' (km²) ~45 (without lake area)
$A':A$ 1.5:1

Geological characteristics

NE: raised bog,
SE: sand of fine grain size

SW: swamp,
NW: sand, partly dunes

Climatic conditions

Average temperature 8.5°C (year)
Average precipitation/year 600 mm
Average sunshine duration
Main wind direction(s) SW(ca 25–30%)
                     W(ca 15–20%)
Evaporation per year ~630 mm (from lake area)

Ice cover (days) 39 (0–ca 90)
Average radiation/year not known
% of calm days not known (ca 10–20%)

Cultural geography and demography

Land usage of catchment area (%)

Industrial

Usage of lake water including recreation activities boating (ca 7000 boats registered, mainly sailing boats), bathing, fishery (commercial and sport)

Agricultural 41
Meadows 23
Forest 19
unused + moor 9
number of residents not known   inhabitants/km² not known

Water temperature: 0.5–1°C min under ice   max ~28°C
Secchi depths: 5–10 cm min   100–120 cm max depending on the amount of
                                        suspended sediment material
Euphotic zone: min   max
O₂ concentration: 70% min   170% max of saturation

pH 6.8–10.2 Conductivity (μS) 280–410   Alkalinity (mval) 0.9–1.8
Average P-conc. ~100–200 mg m⁻³   Average N-conc. ~0.2–3 mg l⁻¹ (inorganic)
            total

Conditions of sediment
  anaerobic, with high water contents (90–95%), high contents of organic matter (40–50% of d.w.), low
  contents of nutrients (especially: Ca^{++} ~0.5–1% Ca/d.w., probably no $CaCO_3$; total P: 0.1% P/d.w.)

Dom. phytoplankton species:
  Microcystis, Anabaena ssp., blue-green algae sometimes dominating under ice: Aphanicomenon (Feb.
  1976), Oscillatoria sp. (Feb. 1979); during summer time: Scenedesmus quadr., Pediastrum boryan

Dom. zooplankton species:
  Keratella cochlearis (seasonal), Bosmina longirostris (perm.), Eubosmina coregoni, Daphnia cucullata

Dom. macrophytes:
  Phragmites communis and Scirpus lacustris (mainly eastern shore line), (Glyceria), submerse flora
  for-reaching resp. nearly totally failing

Dom. benthic organisms:
  very poor, seasonal and regional changing; regional differences at least partly depending on substrate
  conditions (e.g. Mollusca); reasons for seasonal changings not known

Fishes:
  Anguilla anguilla, Esox lucius and Stizostedion lucioperca commercially important; Abramis brama,
  mainly hunger forms (poor benthic fauna), Rutilus rutilus, Acerina cernua more than 50–60% of
  commercial catch by net-baskets

SOME STUDIES ON THE PROBLEM OF "TREIBMUDDE" IN STEINHUDER MEER

J. POLTZ

Abstract

Steinhuder Meer is an extremely shallow lake (mean depth 1.35 m). The autochthonous sediments are frequently suspended by waves and then drifted by currents ("Treibmudde"). It is usually only the upper few millimetres of the mud which are disturbed. The transportation of suspended matter by the outflow is calculated to be about 500 t dry wt. per year. Due to the phenomenon of "Treibmudde" there may be a heavy silting in the wind-sheltered parts of the lake, with up to 25 cm per year being recorded in seston samplers.

There is a rapid decomposition of nitrogen containing compounds in the "Treibmudde" being even detectable in the seston samplers. In contrast the mineralisation of phosphorus compounds is slower. Indirect evidence is presented for a vertical distribution of phosphorus and iron within the sediments with maxima of concentrations at the surface. The calcium contents of the "Treibmudde" and of the sediments are very low, although probably a biogenic precipitation of lime occurs during summer time.

Introduction

The area of Steinhuder Meer is 29.1 km². The lake is the largest shallow water body in the Federal Republic of Germany with a mean depth of 1.35 m (max. 2.80 m). The water is very turbid largely due to suspended sediments. Secchi-disk measurements are 25–30 cm depth on average. Though most of the disturbed mud can quickly resettle the Secchi-disk values are rarely greater than 1 m even during calm periods or under ice. This is due to large crops of planktonic algae, and

probably also some suspended material which has a slow sedimentation rate.

Due to the frequency and velocity of winds in the northern German plain the suspension and transportation of sediments by waves and currents is very common in Steinhuder Meer as well as in the other shallow lakes in Niedersachsen. This is known locally as "Treibmudde" (which may be translated as drifting or floating mud). Related to this phenomenon of "Treibmudde" is the characteristic distribution of sediments in the lake basin, which are mainly deposited in the western part of the lake (Dienemann et al., 1943; Grahle, 1965; Müller, 1968). Moreover, compared with deep eutrophic lakes of similar age, the amount of autochthonous sediments in Steinhuder Meer is rather low, and this also is in some way related to the frequent suspension of mud.

There are two points of interest in the phenomenon of "Treibmudde":

1. Steinhuder Meer is the centre of a recreation area of great importance. Aquatic sports are very popular, especially sailing. It is obvious that the frequent displacement of sediments becomes a nuisance to these users as sedimentation preferentially occurs in places where there is low water movement, e.g. near landing-stages, in channels or passages in the reed belt etc. So understanding the extent of internal movement of sediments is important in developing management policy for the lake.

2. The composition of the "Treibmudde" is of interest in relation to the origin of autochthonous sediments and the extent of nutrient return from sediments to the water in shallow lakes. As shown elsewhere (Poltz, 1978) the sediments of Steinhuder Meer differ from those of comparable deep lakes, especially by

- high content of organic matter (up to, and greater than, 50% of dry weight),
- low phosphorus concentrations,
- very low calcium contents.

Material and methods

The preliminary evaluations presented here include

- 286 analyses of water samples. If the water contained more than 50 mg dry wt. l^{-1} of suspended material (94 samples), it was assumed that the "Treibmudde" predominated over the planktonic biomass (cf. Fig. 1). The estimation of the composition of the suspended material was performed by calculating the regression lines suspended matter/(N-, P-, Fe-) concentration in the lake water (cf. Figs. 2 and 4, Table 1). The relationships were supposed to be linear. In fact they are not linear, because of the different contents in plankton and sediments forming together the suspended material. This supposition, however, is confirmed by high correlation coefficients above 50 mg dry wt. l^{-1}.
- 51 analyses of "Treibmudde" samples, which were collected in funnel-like seston samplers similar to those used by Ohle (1962). The periods of sampling varied from 6 to 22 days depending on the amount of seston yielded.
- 29 analyses of sediment samples from the surface layers of the mud of liquid quality. The thickness of this layer of potential "Treibmudde" is about 3–5 cm having a water content of 92–97%.

All analyses were carried out by the laboratory of the Niedersächsisches Wasseruntersuchungsamt in Hildesheim according to the methods as described in the "Deutsche Einheitsverfahren". The organic material was estimated as ignition loss after heating to 550°C.

Results and discussion

The estimated quantity of suspended material varies from 1 mg dry wt. l^{-1} to 450 mg l^{-1} (ca. 300 samples) or from 9 mg dry wt. l^{-1} to 220 mg l^{-1} (mean values for the whole lake, ca. 50 results), respectively. The amount of 220 mg l^{-1}, which was recorded during a storm, corresponds to a quantity of suspended sediments of about 150,000 m³ per lake area or 5,000 m³ km⁻² and is equivalent to a mud layer of 5 mm depth. So only the upper few millimetres of sediments are usually raised by waves to form the "Treibmudde".

The mean quantity of suspended material for the whole lake is 40 mg dry wt. l^{-1} or for the western part of the lake, near the outflow, 31 mg l^{-1}. Taking this unequal horizontal distribution of suspended material into account, the transportation of particulate matter by the outflow can be calculated as about 500 t dry wt. per year. So the removal of mud may be reckoned to be in the order of about 50% of the autochthonously produced sediments.

The extent of internal movements of sediments can be gauged by the material collected by the seston samplers. According to the actual wind velocity the rate of sedimentation in these samplers accounted for 15 g dry wt. m⁻² d⁻¹ to 130 g m⁻² d⁻¹. There was a clear vertical distribution of drifting mud within the water column. The mean values for the overall sampling period of about 200 days were 35 g dry wt. m⁻² d⁻¹ in 50–70 cm depth and 50 g m⁻² d⁻¹ in 100–120 cm depth, respectively. The latter result corresponds to a rate of sedimentation of about 0.7 mm d⁻¹ or 25 cm per year. Though these data are only true for the artificial conditions in the seston samplers (preventing any resuspension), they give an idea of the possible extent of local sedimentation.

The suspended material and the "Treibmudde" collected by the seston samplers consists of variable amounts of living and dead plankton, of suspended sediments, and another fraction with a low sedimentation rate. The different and changing proportions of these constituents cannot be distinguished. They mainly involve the deviation of the data (see confidence intervals in Fig. 1 and Table 1). So the present interpretation is tentative. It needs, therefore, confirmation by further investigations.

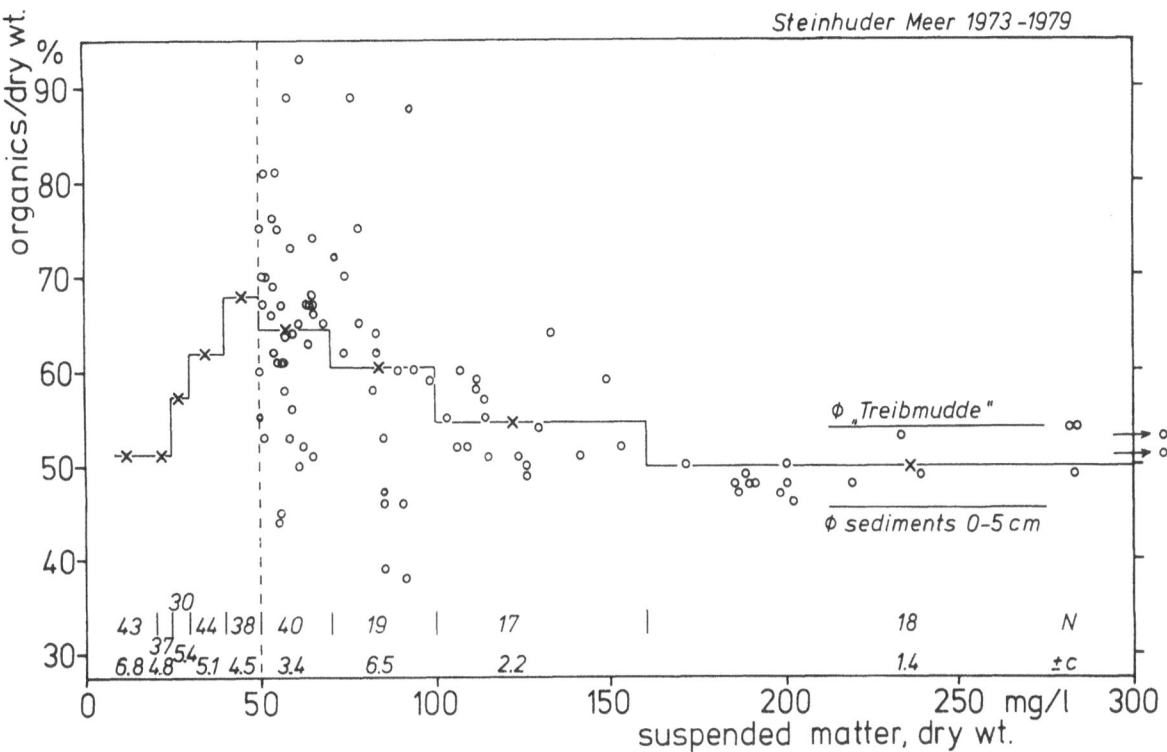

Fig. 1. Variation of the organic content (expressed as a percentage of dry weight) of suspended material (dry weight as mg l^{-1}). The individual results were classified and the mean values of the organic contents calculated (solid lines). The number of analyses of the different classes (N) and the confidence intervals (95%) of the mean values (±c) are given at the lower margin. The dashed line (at 50 mg l^{-1}) marks the limit above which the "Treibmudde" predominates over the planktonic biomass (see text).

The relative proportion of organic matter in the suspended material alters with the amount of particulate matter floating in the water (Fig. 1). Above 50 mg dry wt. l^{-1} it decreases from approximately 65% down to 50% organic content/dry wt. This latter figure is between the mean values for "Treibmudde", as yielded by seston samplers (54%), and the upper sediment layer (45%). This relationship is due to the increasing suspension of sedimentary material with increasing water movement, thus reducing the proportion of planktonic biomass in the suspended matter.

The comparably low organic content (51% of dry wt.) when suspended material values are less than 25 mg dry wt. l^{-1} seems to be remarkable (Fig. 1). This is probably due to the very slow sedimenting material as mentioned above, which seems to be mainly inorganic. With increasing amounts of suspended matter up to 50 mg dry wt. l^{-1} the organic part increases up to 68% of dry wt.,

on average, which is mainly caused by increasing amounts of planktonic biomass.

The Kjeldahl-nitrogen concentration of the suspended material accounts for 3.2% N/dry wt. (Fig. 2), which corresponds to about 6.3% N/organic matter (Table 1). In the "Treibmudde" derived from the seston samplers the N-content is lower. It is about 4.2% N/org. This corresponds to the mean value of the upper sediment layers of the open water area of 4.3% N/org. (Table 1).

The situation found for phosphorus was different. The suspended material and the "Treibmudde" have quite uniform concentrations of 0.24–0.26% P/dry wt. or 0.45–0.48% P/org. These are higher than the corresponding data for the sediments in the open water area of 0.13% P/dry wt. and 0.29% P/org. (Table 1). So probably the P-concentration in the sediments has a heterogenic vertical distribution with a maximum near the surface. This corresponds to the result,

5

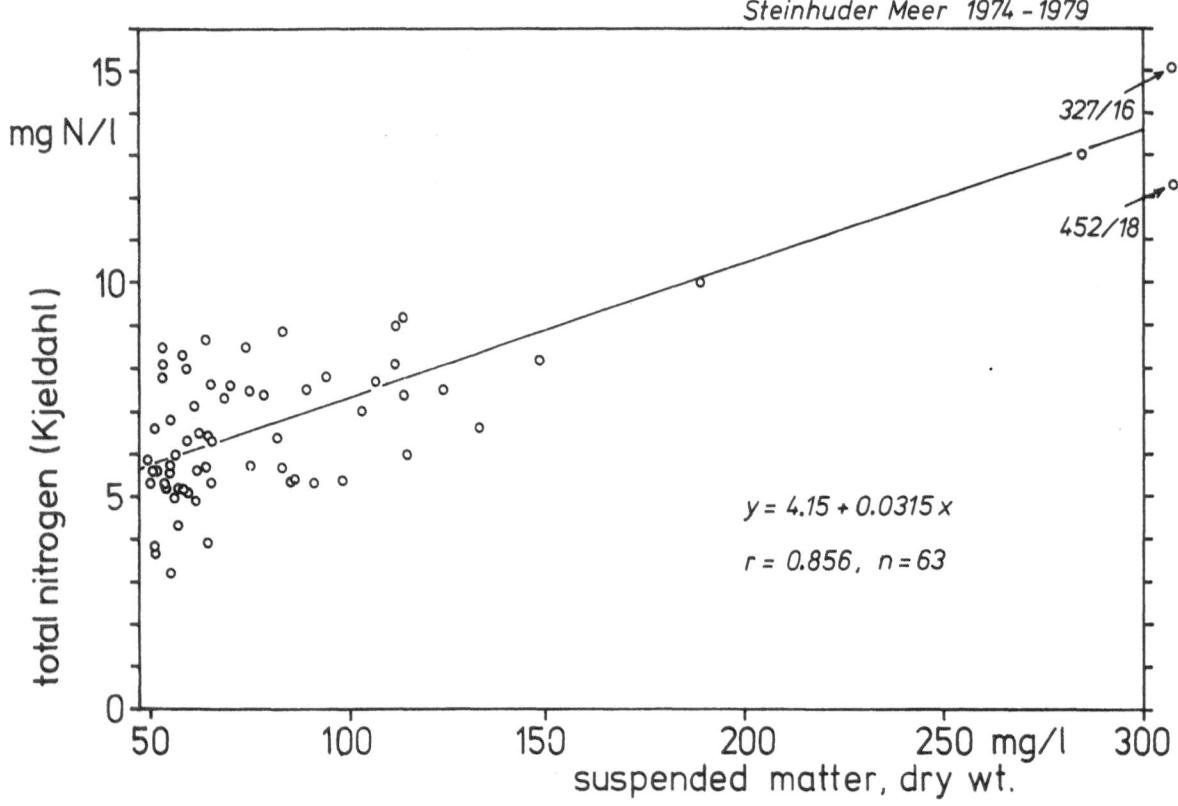

Labels within figure:
mg N/l

total nitrogen (Kjeldahl)

327/16

452/18

$y = 4.15 + 0.0315\,x$

$r = 0.856,\ n = 63$

suspended matter, dry wt.

Fig. 2. Correlation of the amount of suspended material and total nitrogen (dissolved and particulate) in the lake water.

mentioned previously, that it is not the layer of mud down to 3–5 cm (as analysed) that is resuspended, but only the upper few millimetres. Consequently the P-concentration of suspended matter at amounts of more than 300 mg dry wt. l^{-1} (during storms) was lower accounting for 0.19% P/dry wt. or 0.36% P/org.

The different findings for nitrogen and phosphorus may be explained by the different ways and velocities with which they redissolve during the decomposition of organic material. The mineralisation of organic N-compounds is obviously faster than that of P-compounds. It was even detectable in the seston samplers, where the supernatant water may contain more than 20 mg inorganic $N\,l^{-1}$. This is up to a hundred times higher than the concentration in the lake water at the same time. In contrast the P enrichment in the supernatant water of the seston samplers was only 1–5 fold. The comparison of the frequently-suspended-sediments of the open lake area with those of wind-sheltered regions indicates that the

decomposition of N-compounds accelerates under anaerobic conditions. However with the mineralisation of organic P-compounds the frequent disturbance of sediments seems to be more important (Table 1).

The calcium content of the sediments in Steinhuder Meer is very low. It accounts for only 1.2% Ca/inorganic matter in the upper layers (Table 1). Furthermore the ratio Ca:Mg (wt./wt.) in the sediments is 3.8 on average, and it is thus lower than in the lake water (mean 7.4). So it had been concluded previously (Poltz, 1978) that the biogenic precipitation of lime was not important for the formation of autochthonous sediments in Steinhuder Meer. The concentration of dissolved calcium in the lake water, however, is about 42 mg $Ca\,l^{-1}$, and pH-values usually arise to 9–10 during the summer time. So under these conditions a biogenic precipitation of Ca may be expected.

This assumption seems to be confirmed by the data found for the "Treibmudde". This had, on

Table 1. Mean composition and confidence intervals (95%) of suspended material (≥ 50 mg dry wt. l^{-1}), "Treibmudde", and sediments in Steinhuder Meer.

	Suspended material $\geq$50 mg/l	"Treib-mudde"	Sediments 0–5 cm open area	Sediments 0–5 cm sheltered regions
Number of analyses	94	51	21	8
organ. mat. (% of D)	55.3	53.6	45.4	37.8
	±2.1	±1.7	±2.3	±4.4
total N (g kg^{-1} D)	31.5[1]	22.7	19.2	13.8
	±4.8	±2.3	±2.8	±4.5
(g Kg^{-1} O)	63.2[1]	41.9	43.0	32.8
	±9.4	±3.8	±5.3	±11.5
total P (g Kg^{-1} D)	2.58[1]	2.35	1.30	1.85
	±0.42	±0.29	±0.21	±0.45
(g Kg^{-1} O)	4.84[1]	4.46	2.92	4.92
	±0.99	±0.57	±0.43	±1.26
Ca (g Kg^{-1} I)	—	17.9	11.9	6.9
	—	±2.6	±3.1	±2.8
Ca:Mg (wt. wt^{-1}.)	—	4.5	3.8	2.4
	—	±0.7	±0.8	±1.2
Fe (g Kg^{-1} I)	62.5[1]	58.6	47.2	30.6
	±3.2	±4.5	±6.1	±4.0
Mn (g Kg^{-1} I)	—	2.34	0.98	0.57
	—	±0.67	±0.13	±0.15
Fe:Mn (wt. wt^{-1}.)	—	26	51	57
	—	±1.7	±5	±12

D = dry weight. O = organic matter. I = inorganic material.
[1] Values derived from the inclination of the calculated regression lines of suspended material/concentrations (N P, Fe) in the lake water (cf. Figs. 2 and 4).

average, a content of 1.8% Ca/inorganic material (max. 3.6%) and with a ratio of Ca:Mg of 4.5 (max. 9.7). This is also indicated by the comparison of Ca-contents and Ca:Mg ratio of the "Treibmudde", on the one hand, and the amount of "Treibmudde" collected in the samplers and pH of the lake water on the other (Fig. 3). The curves for Ca-concentration and Ca:Mg ratio follow a similar trend. They are inversely related to the lines showing the relative amount of "Treibmudde" trapped per unit time. Moreover, they increase with increasing pH of the lake water.

The Ca-concentration and the Ca:Mg ratio of the "Treibmudde" are, however, very low compared to data for sediments of the gyttja type cited in literature (e.g. Ungemach, 1960). So it is still unclear whether there is only a little or even no Ca precipitation or whether the precipitated lime is rapidly redissolved due to the speed of mineral-isation of the organic material. The Ca-concentrations of the supernatant water in the seston samplers, however, was only 10–20% higher than in the lake water. So there is not as rapid a redissolving of Ca as that found for nitrogen.

The concentrations of heavy metals decreased in the following sequence (1) suspended material (2) "Treibmudde" (3) sediments (Table 1), especially for manganese relative to iron. This is in agreement with the well known faster redissolving of Mn from mud (cf. Hutchinson, 1957). Thus the ratio Fe:Mn (wt./wt.) increases in the same succession. The differences in the Fe-contents of suspended material, "Treibmudde", and sediments, however, was less than found for nitrogen and phosphorus. So there is a good linear correlation of the amount of suspended matter and the concentration of total iron in the lake water (Fig. 4).

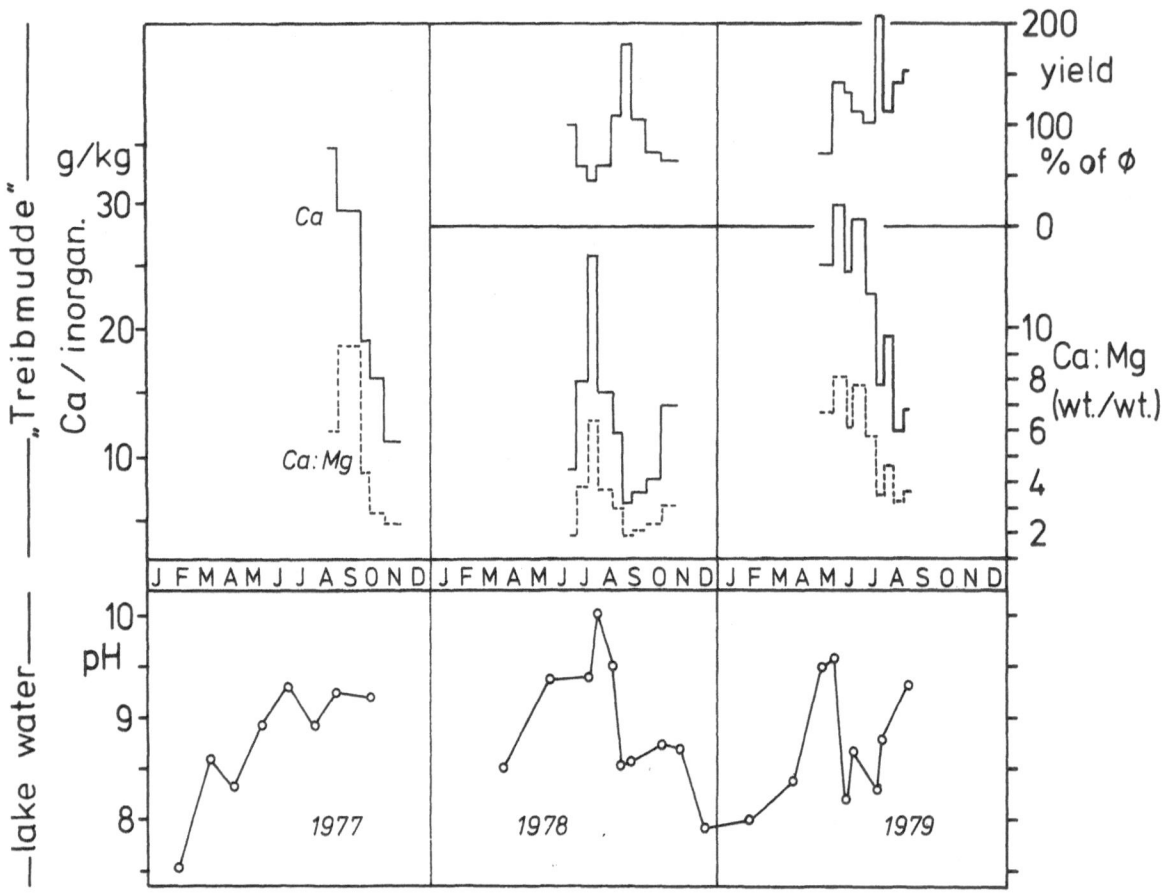

Fig. 3. Amount of "Treibmudde" collected in the seston samplers per unit time (yield as % of average), calcium concentration and Ca:Mg ratio of the "Treibmudde", and pH of the lake water.

According to the data given in Table 1, however, one should expect the Fe-content of the suspended matter to decrease as more sediments are raised by waves, but obviously this is not true. So, similar to the conclusion for phosphorus, it is to be supposed that there is a vertical gradient of Fe-concentration within the upper sediments which has its maximum near the surface. This is indeed not unusual for lake sediments (cf. e.g. Groth, 1971; Tessenow, 1972) according to the redox relationship in the sediments and at the mud-water-interface.

Though there is a similar vertical distribution of iron and phosphorus within the sediments of Steinhuder Meer (which has, however, still to be proved directly), the previous results do not indicate a direct relationship between Fe and P. This can be seen from a comparison of the data for the frequently suspended sediments of the open lake area on the one hand and those of the sediments of the sheltered regions on the other (Table 1). In the latter case the mean P-concentration is higher than in the former but the Fe-content lower.

In summary, the results of our preliminary investigations are hardly more than a description of the situation of Steinhuder Meer. Moreover, they show the individual properties of this shallow lake rather than pointing out general principles. This may be illustrated by some examples such as in Bederkesaer See where the phenomenon of "Treibmudde" is of great importance for the nutrient budget, as the dissolved phosphorus is bound and precipitated by suspended sediments (Poltz, 1977). In contrast to Steinhuder Meer the sediments of the shallow lake Dümmer contain up to 15% lime per dry weight (estimated by gas volumetric analysis of carbonate, Dahms, 1972) or even up to more than 12% Ca/dry wt. (unpubl.

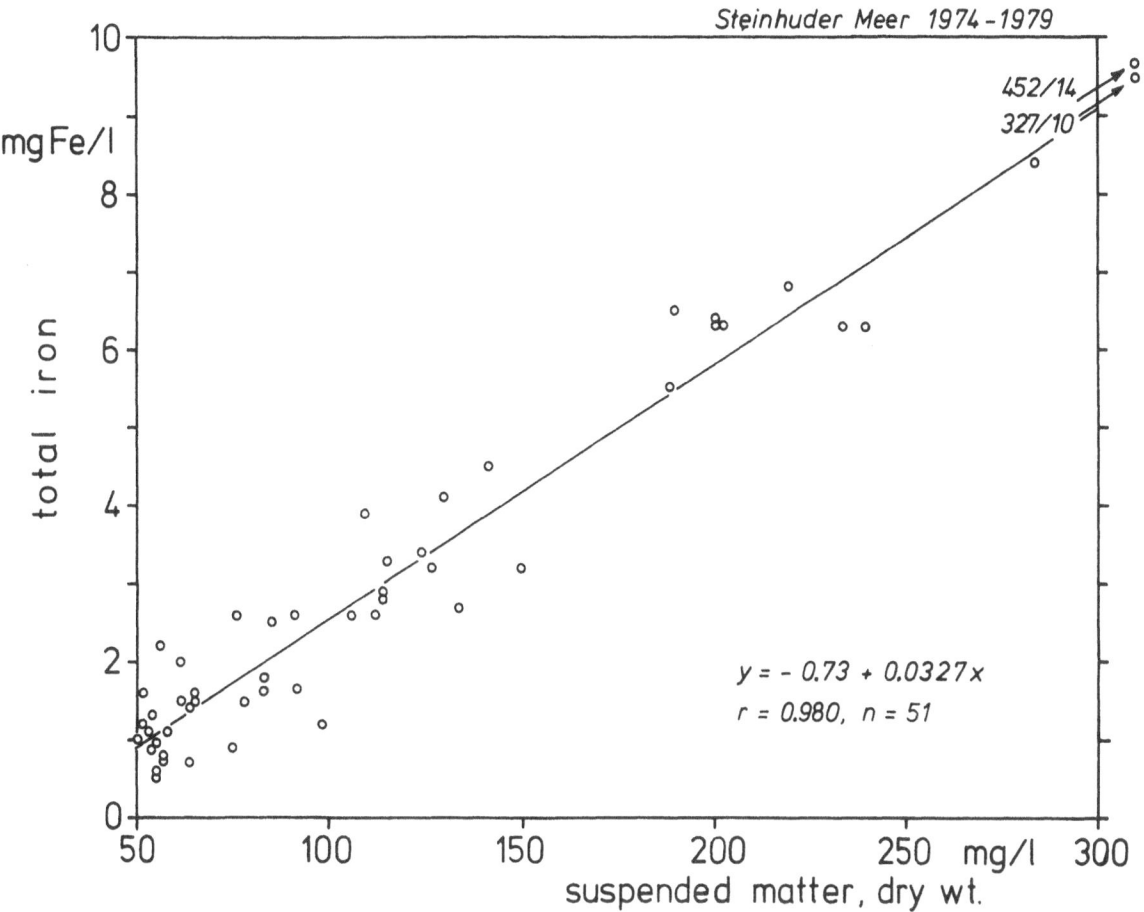

Fig. 4. Correlation of the amount of suspended material and total iron (dissolved and particulate) in the lake water.

data). And finally, as pointed out in contributions and discussions of this symposium, on Neusiedlersee (Neuhuber) and Lake Balaton (Entz) there are very few or hardly any autochthonous sediments to be found in the free water area. So the phenomenon and problem of "Treibmudde" in Steinhuder Meer is not comparable with the situation found in these two lakes.

References

Dahms, E. 1972. Limnogeologische Untersuchungen im Dümmerbecken im Hinblick auf seine Bedeutung als Landschafts- und Naturschutzgebiet. Thesis, Berlin.

Deutsche Einheitsverfahren zur Wasser-, Abwasser- und Schlammuntersuchung 1960. Ed.: Fachgruppe Wasserchemie (GDCh), Weinheim.

Dienemann. W. & Pfaffenberg, K. 1943. Zur Alluvialgeologie des Steinhuder Meeres und seiner Umgebung. Arch. f. Nds. 19: 430–448.

Grahle, H. O. 1965. Bericht über die limnogeologische Untersuchung des Steinhuder Meeres (Geol. Unters. an nds. Binnengew. IV). Unpubl. report of the Nds. Landesamt f. Bodenforsch., Hannover.

Groth, P. 1971. Untersuchungen über einige Spurenelemente in Seen. Arch. Hydrobiol. 68: 305–375.

Hutchinson, G. E. 1977: A treatise on limnology, vol. I. New York, London.

Müller, H. 1968. Zur Entstehung und Entwicklung des Steinhuder Meeres. GWF Wasser Abwasser 109: 538–541.

Ohle, W. 1962. Der Stoffhaushalt der Seen als Grundlage einer allgemeinen Stoffwechseldynamik der Gewässer. Kieler Meeresforsch. 18: 107–120.

Poltz, J. 1977. Observations concerning the nutrient exchange between water and sediments in some eutrophicated shallow waters in Niedersachsen. Proc. Internat. Sympos. EUTROSYM 76, vol. III: 273–281, Karl-Marx-Stadt, Berlin.

Poltz, J. 1978. Untersuchungen an Sedimenten in niedersächsischen Flachseen I. Arch. Hydrobiol. 82: 1–19.

Tessenow, U. 1972. Lösungs-, Diffusions- und Sorptionsprozesse in der Oberschicht von Seesedimenten I. Arch. Hydrobiol. Suppl. 38: 353–398.

Ungemach, H. 1960. Sedimentchemismus und seine Beziehung zum Stoffhaushalt in 40 europäischen Seen. Thesis, Kiel.

PRIMARY PRODUCTIVITY MEASUREMENTS AND CARBON METABOLISM IN STEINHUDER MEER AND LAKE DÜMMER

D. ERNST & D. REINHARDT

Abstract

Steinhuder Meer and Dümmer are the two largest lakes in Niedersachsen (northern Germany). They are similar in climate and origin. They differ in allochthonous input. Measurements of primary production (^{14}C-method) and several chemical and meteorological parameters have been carried out since 1974. The trophic states are compared to other lakes on the basis of annual production per unit area, production density, maximum of production, NO_3 and PO_4. Both lakes appear polytrophic. Primary production, CO_2-input from the atmosphere, C-output via the outflow and $CaCO_3$ precipitation are estimated for Steinhuder Meer.

Introduction

Steinhuder Meer and Lake Dümmer are the two largest lakes in Niedersachsen (northern Germany). Both lakes are major recreational areas and are important nature reserves and fishing sites. The Dümmer, which is surrounded by a dam, also serves as a reservoir to prevent flooding. Both lakes are surrounded by reed belts and are similar in size, depth, climate, exposure to wind, precipitation and origin.

Dienemann (1963) has suggested the origin was as a result of deflation, although more recently drilling results by Dahms (1974) claim a thermocarst origin for Lake Dümmer and probably also for Steinhuder Meer.

The lakes differ in their eutrophic status. There is a heavy nutrient loading from the river (the Hunte) flowing into the Dümmer, whereas the Steinhuder Meer is almost exclusively fed by rain and subsoil water. The most recent paper on the hydrology of the latter is by Plate (1975). Both lakes are very shallow and do not display thermal stratification.

In Steinhuder Meer the subsoil is of sand covered with a thin mud layer which increases towards the west. Age and thickness of the mud- and peat-layers of the west-basin have been investigated by Müller (1969) and the components of the recent sediments in relation to lake metabolism have been discussed by Poltz (1978).

The top layer of the black mud is easily resuspended by wind and causes a high turbidity which restricts submerged macrophytes and the fish feeding of bottom fauna. The euphotic zone may be as shallow as 1 m.

Primary productivity is an important eutrophication parameter particularly as it is more or less the only one which applies to shallow lakes. The Institute for Biophysics, Hannover, began in 1974 in Steinhuder Meer and in 1975 in Lake Dümmer to measure this as well as chemical and meteorological parameters in these lakes. In this paper we present the results and interpret them in relation to the carbon metabolism, particularly for Steinhuder Meer.

The lack of stratification and the high turbidity (typical properties of shallow lakes) are discussed in relation to their influence on primary productivity.

Methods

Primary productivity, ^{14}C-method

The primary production was measured by ^{14}C-uptake following the methods of Steemann-Nielsen (1952) and Vollenweider (1974). The support of the bottles during exposure has been designed carefully to prevent shading (Plexiglas supports with very little metal, double body buoys) and to cover the dark bottles (black plastic shrunk cover). Selecting the proper exposure time requires a compromise between the errors by bottle effects over long times and the influence of manipulation times and counting statistics which dominate at short times. Exposure periods between 5 min and 12 hr have been checked with 3–4 h giving the least errors. The activity added was 0.3–1 μCi per bottle; the bottle size was about 100 ml. Samples were filtered on to membrane filters (Sartorius No. SM 11106, 23 mm $\varnothing$, 0.45 μm pore diameter) with 0.5 atmospheres of vacuum. Radioactivity was measured by liquid scintillation counting (scintillator: toluol, ethanol, PPO, POPOP, and ethanolamine) of wet filters after Wallen and Geen (1968), a method which prevents losses of [14]CO_2 on drying. Quenching was checked by the measurements of series of decreasing algal concentrations. Trials to dissolve the filter in the scintillator (e.g. with "soluene" (Packard)) resulted in high quenching, so did cellulose nitrate filters. Horizontal, diurnal and depth-distribution has been checked repeatedly. Routine measurements were carried out every 2–3 weeks in the first year, but less often in subsequent years.

The inorganic carbon was determined by a modified "p" and "m"-value method (Titration at pH 8.2 and another value which is individually determined for each sample) after Goltermann (1969). The determination of dissolved inorganic carbon (DIC) usually causes the most prominent error in such experiments. It could be reduced to about 5%. The experimental errors sum up to 25% for absolute and 10% for relative determinations.

The ^{14}C uptake results have been checked with the oxygen evolution method. An exact comparison requires the knowledge of the respiration-coefficient. Applying a value of 1.1 the two methods agree exactly. In the literature usually slightly higher values are reported but are within the range of experimental errors.

Chemical parameters

During the field experiments pH, conductivity, O_2-content, temperature, Secchi disk reading and wind velocity have been determined. Further chemical analyses – mainly after Deutsche Einheitsverfahren (1975) – cover: monophosphate, total phosphate, $KMnO_4$-consumption, NH_3, NO_3, NO_2, Ca, Mg, Fe, as well as occasionally Cl, SO_4, Na, K, H_2S.

Meteorological parameters

Meteorological parameters were at first measured in a small meteorological station near the lake. In 1976 a floating limnometeorological station was built which collected all the relevant data: air- and water-temperature, wind-velocity and -direction, global radiation, under-water light and turbidity.

Results

The results of productivity measurements and most of the other important parameters are shown in Fig. 1a and 1b.

The diurnal production curve is, within experimental errors, symmetrical about midday (Fig. 2) in Steinhuder Meer (Reinhardt, 1977).

The production per unit area – i.e. the depth integral of production – has been correlated to the production at 20 cm depth. The correlation is significant ($r = 0.814$, $n = 16$) and allows the possibility of calculating the depth integral from just one ^{14}C-measurement and the Secchi value (Fig. 3) (Reinhardt, 1977).

$$P_{tot} = \frac{P_{20}}{3.22 - 0.0321 \cdot s}$$

P_{tot}: (mg C m^{-2} h^{-1}) – depth integral of production
P_{20}: (mg C m^{-3} h^{-1}) – production in 20 cm depth
s: (cm) – Secchi-reading

Assimilation is restricted to the upper 100 cm and displays a maximum (Fig. 2b, 4).

A trial to get a correlation between Secchi-reading and mass of suspended mud failed. Mid

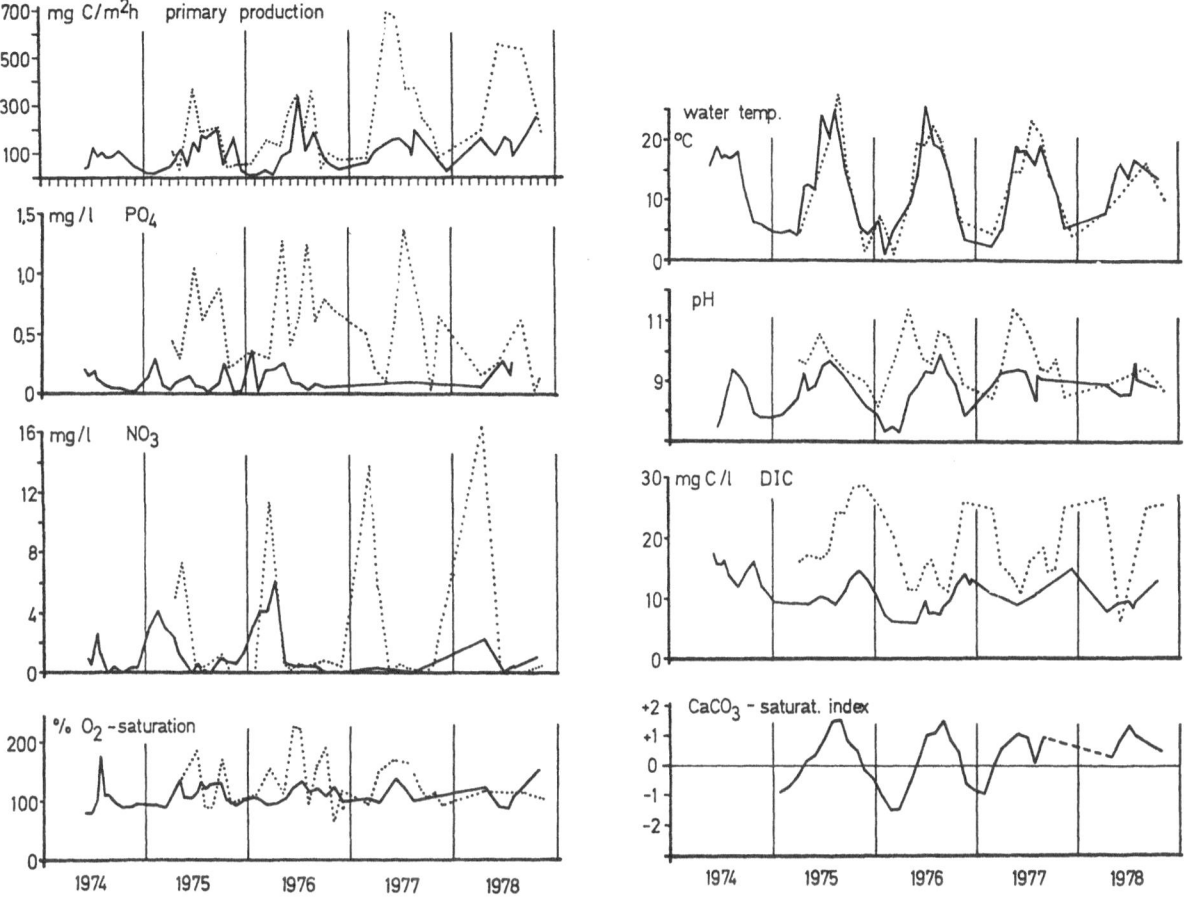

Fig. 1a and b. Development of the most important limnological parameters from 1974 to 1978
in Steinhuder Meer (————) and Dümmer (. . . . . .).

lake samples under average or calm weather conditions gave a poor correlation ($r = -0.665$). The correlation coefficient over all weather conditions and stations ($r = -0.329$) is even worse (Fig. 7). Apparently grain size and composition of the mud varies considerably. However, the data collected for this purpose gave a crude estimate of how much organic matter may be suspended and carried away with the outflow.

From the production- and radiation data one may calculate the "efficiency" of this ecosystem (enthalpy of the produced organic matter over incident light energy). The values vary between 0.5 and 0.1% depending on weather conditions. These figures are well within the range of other fresh water ecosystems. A statistical analysis re-

veals a result which is well expected: the efficiency increases with temperature and decreases with total underwater **irradience (Fig. 5)**.

Discussion

Comparison of eutrophication parameters to other lakes

Vollenweider (1968) gives a comprehensive revue of eutrophication parameters. For each parameter he gives ranges for the trophic states. In Table 1 an outline of these figures is given together with the corresponding values for the two lakes under investigation here. It must be mentioned that the reference figures all refer to deep lakes. Judged by

13

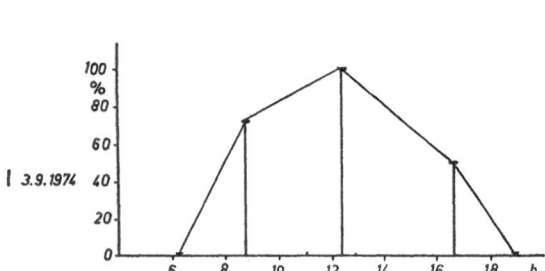

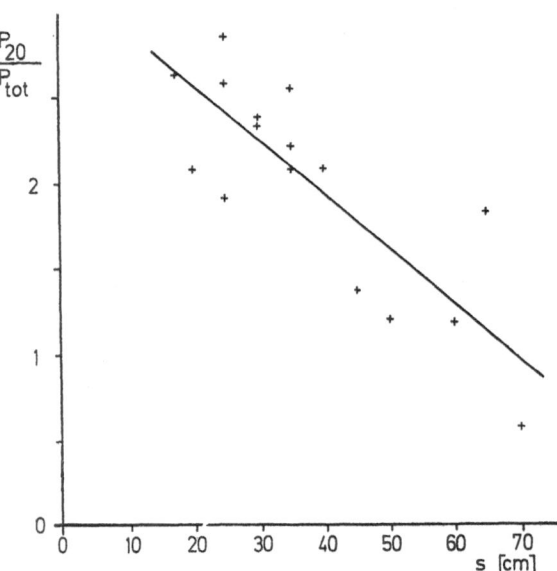

Fig. 2. Symmetry of diurnal development of primary production. (a) Production per unit area integrated over depth as a function of time. (b) Production as function of light intensity from two depth profiles, one in the morning and one in the afternoon. They do not show differences.

Fig. 3. Correlation between Secchi-reading (s(cm)) and P_{20}/P_{tot} (see Text) in Steinhuder Meer.

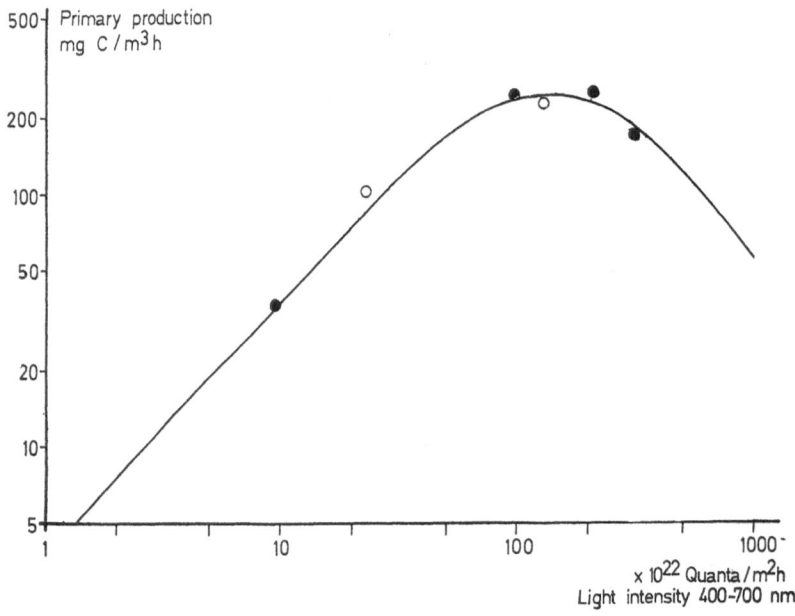

Fig. 2b.

all these parameters both lakes have to be rated polytrophic. The Dümmer being more heavily charged by allochtonous nutrients. Particularly the production data show an enormous turnover. Due to the shallowness the oxygen supply is excellent, the O_2-saturation always stays around 100%. The seasonal variations of pH, NO_3 and PO_4 would require further interpretation. However our present knowledge of the circumstances is insufficient to give a final discussion. In the Steinhuder Meer interpretations of nutrient concentrations are impossible as long as there are no analyses of the subsoil inflow which supplies one half of its water. In the Dümmer which has a retention time of only

14

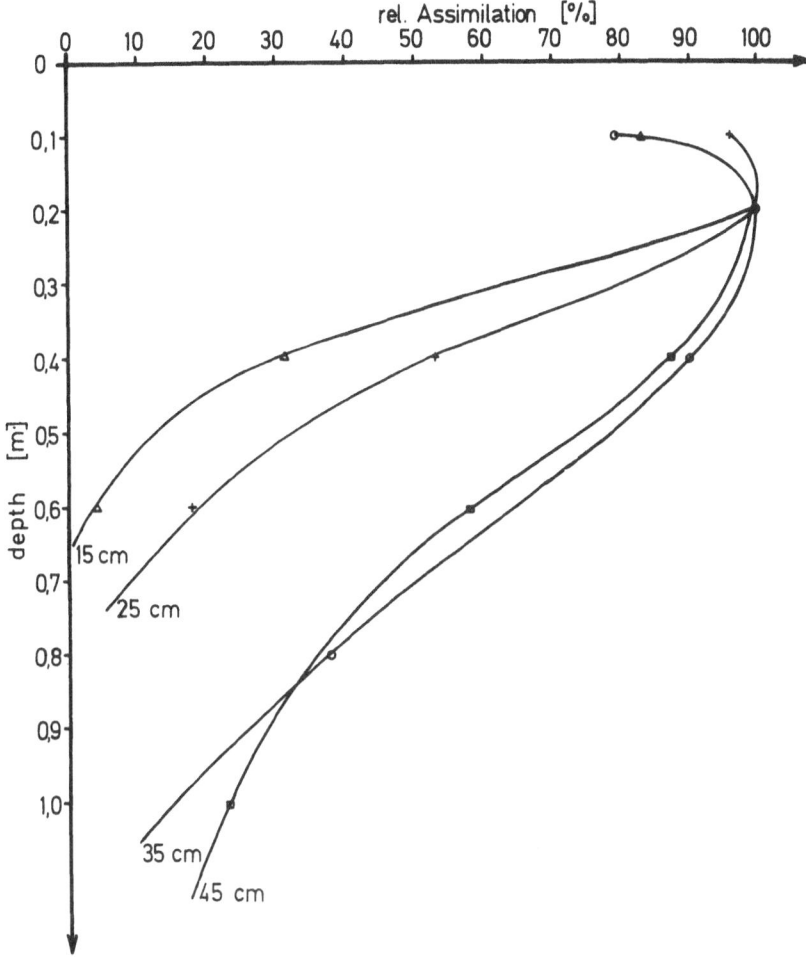

Fig. 4. Four examples for assimilation (%) as function of depth (m) on 4 selected days with different Secchi-readings (cm). The value at 20 cm depth was set to 100%.

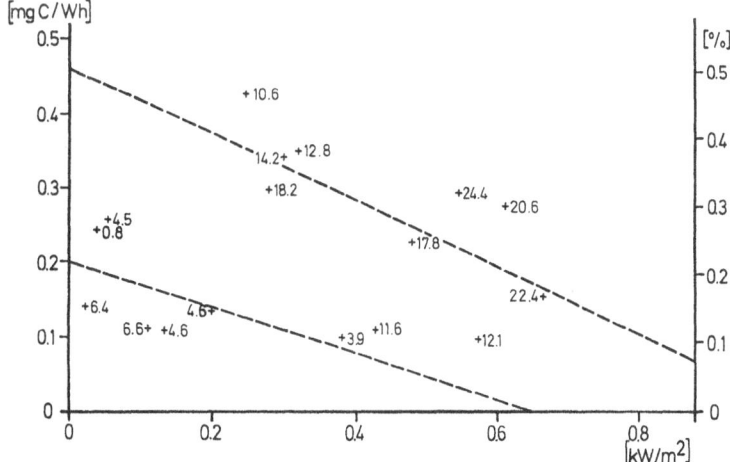

Fig. 5. Assimilation (mg C/Wh) and efficiency (%) as function of total incident light (kW/m²) for two temperature-ranges: >10°C, upper line, correlation $r = 0.66$ and <10°C lower line, $r = 0.59$.

Table 1. Classification of Dümmer and Steinhuder Meer with respect to trophic state (comparative data for Schöhsee a.s.o. by Ohle (1962) others by Vollenweider (1968)).

1) annual production per area

oligotroph		50 g C/m^2a
polytroph		150 or 200 g C/m^2a
Steinhuder Meer	1975	306 g C/m^2a
	1976	280 g C/m^2a
Dümmer	1975	416 g C/m^2a
	1976	592 g C/m^2a

2) production density averaged over euphotic zone and over time (May-Dec.)

			euphotic zone
Schöhsee		24,2 mg C/m^3d	13,5 m
Schluensee		19 mg C/m^3d	17 m
Plußsee		210 mg C/m^3d	5 m
Steinhuder Meer	1975	1001 mg C/m^3d	1 m
	1976	1806 mg C/m^3d	1 m

3) Photosynthesis at the highest productive depth per day limit for eu-or

polytrophic state	0,2 or 0,3 g C/m^3d (refering to deep lakes)
Steinhuder Meer	4,0g C/m^3d
Dümmer	11,0 g C/m^3d

4) NO_3-concentration

polytrophic state		1500 mg N/m^3
Steinhuder Meer mean		
	1975	1700 mg N/m^3
	1976	2100 mg N/m^3

5) PO_4-concentration

meso-eutrop		10-30 mg P/m^3
eu-polytroph		30-100mg P/m^3
polytroph		100mg P/m^3
Steinhuder Meer mean		
	1975	52mg P/m^3
	1976	63mg P/m^3

3–5 weeks the concentrations are largely influenced by the circumstances in the watershed of the river Hunte.

The apparent pH peaks in summer are due to the strong CO_2-consumption which increases pH particularly as the water is poorly buffered.

Some results concerning the C-metabolism of Steinhuder Meer

It has been impossible up to now to give the detailed carbon balance as two values were not known:

1) the C-content of the underground influx
2) the CO_2-drain into the atmosphere.

However some data are known and shall be reported here.

If we consider carbon data averaged over the year the picture is as follows: From primary production $239\,g\,C\,m^{-3}$ ($300\,g\,C\,m^{-2}\,a^{-1}$) are fixed. The outflow carries 7.2% of this carbon away (DIC: 2%, partic. inorg. C: 0%, diss. org. C: 3.6%, partic. org. C: 1.6%, data from Fig. 1 and 7 and unpublished data by Wasseruntersuchungsamt Hildesheim).

Rain should not transport significant amounts of carbon.

The input of organic material from the reed belts has been checked in another research program and was found to be negligible. (Ernst, D., unpublished data.)

There is a CO_2-exchange with the atmosphere. The saturation concentrations for water in contact with air have been compared to the actual concentrations of dissolved inorganic carbon (DIC): Usually CO_2 is taken up by the lake except in a few winter months (1974/75: Oct./Jan., 1975/76: Dec./Mar., 1976/77: Nov./Jan.) where the flow is reversed.

The quantitative estimation of such a CO_2-flow on a solely theoretical basis is very unreliable (Verduin, 1975; Schindler, 1975; Wood, 1974). However there are data from tank experiments (originally designed to check nutrient limitation) which allow a crude determination of the CO_2 inflow. These are based on ^{14}C-measurements car-

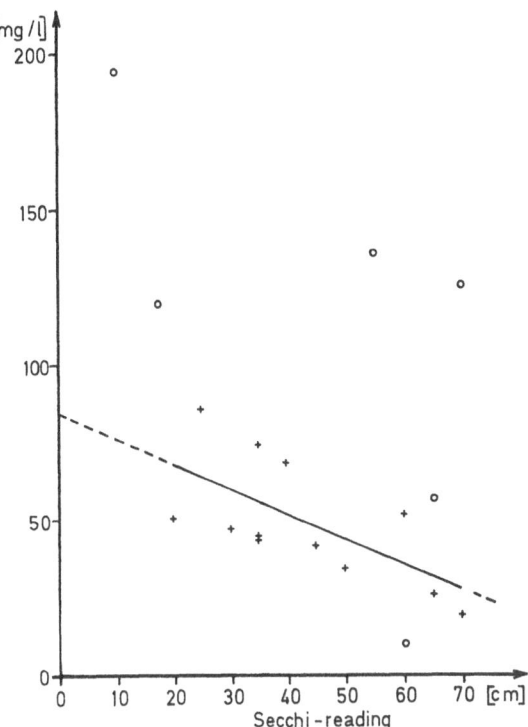

Fig. 7. Mass of suspended mud (dry matter) as function of Secchi-reading + mid-lake samples on calm or average days, O mid-lake samples on windy days and data from other stations.

ried out in tanks and in the open lake. As soon as the water is enclosed in the tank and thus separated from the bottom sediments but still in contact with the atmosphere the DIC-concentration decreases relative to the open water (e.g. Fig. 6). By comparing the variation in DIC with the primary production one can estimate what part of the consumed DIC is coming from the atmosphere. The results were: 37% (Sept. 1977), 10% (Apr. 1978) and 38% (May 1979).

This means that about one third of the carbonate demand, at least during the summer months, is supplied by the atmosphere. The rest originates largely by mineralisation of suspended matter, particularly in the upper layers of the sediment.

This view is supported by an observation during the 1979 experiment: A storm which came up during the work increased the DIC-concentration from 3.2 to 4.2 mg C/l apparently by stirring the upper sediment layer and pH fell from 10.1 to 9.7.

Calcium carbonate precipitation solution can be

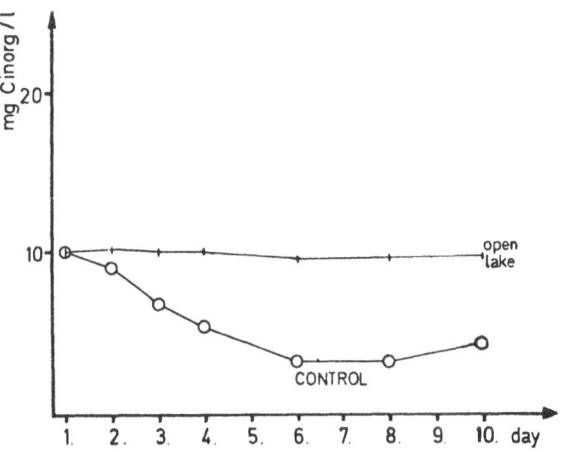

Fig. 6. Decrease of dissolved inorganic carbon as function of time in plastic tank and open lake.

17

checked by calculating the saturation index from Ca and carbonate concentrations, pH and temperature. During the summer carbonate is precipatated (1975: Apr. to Oct. 1976: May to Oct.) but during winter apparently redissolved (Fig. 1b). The interstitial water of the sediment is always acid enough to prevent precipitation. Thus an accumulation of inorganic carbon in the sediments can be ruled out. This is supported by Poltz (1978).

A significant deposition of organic carbon is also ruled out by Poltz (1978) who gives evidence that most of the organic material is remineralised.

There is a seasonal maximum of DIC in fall (Fig. 1b). It may be due to an increase of mineralisation over fixation or to a change in underground carbonate supply. Anyway this increase accounts for only 4% of the annual production.

Aknowledgements

The Wasseruntersuchungsamt Hildesheim (Dr. Poltz) and the Medizinaluntersechungsamt Hannover (Dr. Höppken, Dr. Mühlenberg) have supported this work by carrying out numerous analyses and by many helpful discussions. This work was supported by the International Atomic Energy Agency (research contract 1723 GFR) and the Ministerium für Wissenschaft und Kunst Hannover, via financial support from the Zahlenlotto.

References

Dahms, E. 1974. Geologische und limnologische Untersuchungen zur Entstehungs- und Entwicklungsgeschichte des Dümmer. Ber. Naturhist. Ges. 118: 7–65.

Deutsche Einheitsverfahren, 3. Aufl. 1975. Verl. Chemie, Weinheim/Bergstr.

Dienemann, W. 1963. Zur Entstehung des Steinhuder Meeres und des Dümmers. Neues Archiv für Niedersachsen 12: 230–249.

Goltermann, H. C. & Clymo, R. S. 1969. Methods for Chemical Analyses for Freshwaters. IBP Handbook No. 8, Blackwell Scient. Public., Oxford.

Müller, E. 1969. Diskordanzen und Umlagerungserscheinungen in holozänen Sedimenten flacher Seen Nordwestdeutschlands. Mitt. Internat. Verein. Limnol. 17: 211–218.

Ohle, W. 1962. Der Stoffhaushalt der Seen als Grundlage einer allgemeinen Stoffwechseldynamik der Gewässer. Kieler Meeresforschungen 18: 107–120.

Plate, V. 1975. Hydrologische Untersuchungen am Steinhuder Meer. Wasserwirtschaftsamt Hannover.

Poltz, J. 1978. Untersuchungen an Sedimenten in niedersächsischen Flachseen. I Die Sedimente des Steinhuder Meeres. Arch. Hydrobiol. 82: 1–19.

Reinhardt, D. 1977. Die Messung der Globalstrahlung und Unterwasserlicht und ihr Einfluß auf Primärproduktion in zwei niedersächsischen Flachseen. Diplomarbeit, Uni Hannover.

Schindler, D. W. 1975. Factors affecting gas exchange in natural waters. Limnol. Oceanogr. 20: 1053–1055.

Steemann-Nielsen, E. 1952. The use of Radio-Active Carbon (^{14}C) for Measuring Organic Production in the Sea. J. Cons. Int. Explor. Mer. 18: 117–140.

Verduin, J. 1975. Rate of Carbon-dioxide Transport across Air-Water Boundaries in Lakes. Limnol. Oceanogr. 20: 1052–1053.

Vollenweider, R. A. 1968. Scientific Fundamentals of the Eutrophication of Lakes and Flowing Waters. OECD Technical Report, Paris.

Vollenweider, R. A. 1974. IBP Handbook No. 12. Primary Production in Aquatic Environments. Blackwell, Oxford.

Wallen, D. G. & Geen, G. H. 1968. Loss of Radioactivity During Storage of ^{14}C-labelled Phytoplancton on Membrane Filters. J. Fish. Res. Bord. 25 (10): 2019–2024.

Wille, W., Dembke, K. & Poltz, J. 1976. Limnologische Untersuchungen des Steinhuder Meeres 1964–1971. Niedersächsisches Wasseruntersuchungsamt, Hildesheim.

Wood, K. G. 1974. Carbon-dixide Diffusivity across the Air-Water Interface. Arch. Hydrobiol. 73: 57–69.

PRIMARY PRODUCTIVITY, LIGHT MEASUREMENTS, AND MATHEMATICAL INTERPRETATION

D. REINHARDT

Abstract

A detector system was designed for continuous registration of photosynthetically available radiation under water. Mathematical models describing the relation between light and photosynthesis and for integrating photosynthesis under a unit surface area are discussed and applied to experimental data.

Introduction

Primary productivity measurements at monthly intervals give only instantaneous rates. The conversion to annual production is frequently done by multiplication with a mean day length and then linear interpolation between the monthly measurements. However, in this way one does not account for the daily changes in productivity caused by changing light conditions. It would be ideal to have a continuous measurement and recording of primary productivity. This could be done by using an automatic instrument like the apparatus described by Levin and Lindgren (dates not given). A much easier way would be the continuous registration of changing underwater light conditions by measuring photosynthetic active radiation and vertical attenuation of light. Primary productivity could then be calculated by using an appropriate mathematical formulation describing the relation between light intensity and productivity. This assumes that changes in other likely controlling factors such as species composi-

tion, water chemistry, temperature, etc. are amply monitored by the chosen sample intervals.

Methods

Primary productivity measurements have been carried out on the Steinhuder Meer since spring 1974. Measurements were made about every 4 to 6 weeks. Samples were incubated in duplicate at 10, 20, 40, and 60 cm depth, with 2 additional dark bottles at 20 cm. Exposure time was 3 to 4 hours at midday. Photosynthetic rates were measured by C-14 uptake, with the filters prepared after Wallen and Geen (1968) and counted in a Bertold liquid scintillation counter. Inorganic carbon was determined after Golterman and Clymo (1969).

Underwater light measurement

Quantum efficiency of photosynthesis varies with wavelength. Fig. 1 shows two examples of different efficiency spectra for the same alga species (after Heath, 1972 and Bjoern, 1975). Numerous examples of different efficiency spectra can be found in literature, depending on species and conditions of cultivation. DIN 5031 (prelim.) defines an efficiency curve for photosynthesis to be used with detectors for photosynthetic active radiation. Until recently the most common method of light measurement was using selenium cells, which have a peak response to incident light at about 550 nm,

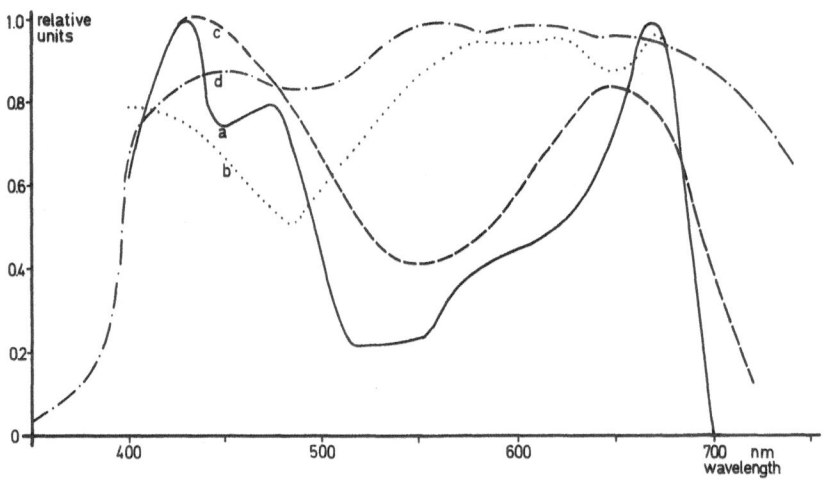

Fig. 1. a, b, action spectra of Chlorella after different authors (a taken from Heath 1972, b taken from Bjoern 1975). c. quantum efficiency of photosynthesis (DIN 5031 prelim.) d. sensitivity of underwater quantum photometer.

sometimes in combination with coloured glass filters (Sauberer, 1941, 1959, 1962; Vollenweider, 1969; Weinberg, 1974). We decided to design an instrument which gives an output signal proportional to quantum flux in the range of 400–700 nm. This conforms to the concept which is now recommended by several authors (Heath, 1972; Lewis, 1975). Our quantum photometer achieves a flat response to within ±8% from 400–700 nm.

Usually light intensity is measured by a flat detector with a cosine response. Since algae suspended in water do not have any preferred direction, correct light measurements require a detector with 4π response. Such an instrument, using scattering spheres made of teflon, was recently described by Booth (1976). The light scattered back and reflected from the bottom (e.g. in extremely shallow waters) usually can be neglected, therefore a detector with 2π response seems to be sufficient. Instruments with a more or less perfect 2π response were described by Heath (1972) and Weinberg (1974). Both authors used scattering domes made of plastic material.

Our experiments showed that plastic domes cannot be used suspended in water for extended periods without degradation of mechanical and optical properties. Therefore we designed a quantum photometer with a glass dome which showed

no detectable deterioration after two years of continuous usage. Only occasional cleaning by wiping with a piece of cloth was necessary. The glass dome consists of a hemisphere of 24 mm outer diameter on top of a short cylindrical part. Both parts are coated with a scattering material on the inner surface. The whole instrument is designed in such a way that a 2π response is achieved with a theoretical error of less than ±1%. Due to some manufacturing problems the angular variation of sensitivity of the assembled quantum detector amounts to about ±6%.

Two of the detectors described above are fixed to a floating platform at depths of 10 cm and 30 cm. The platform is connected to a land based recording station via a 500 m cable. The signal from the upper light detector is integrated for a period of 2 hours and continuously recorded on a paper chart recorder. After every two hours the signal corresponds to the total number of quanta within that period. Measurement of the light integral avoids the difficulties in reading continuous records at times when there are rapidly changing light conditions. It also facilitates calculation of diurnal irradiation. The signal from the lower detector is divided by the signal from the upper detector to give the transmission of a 20 cm water column in the natural light field. Because of the angular distribution of the natural underwater

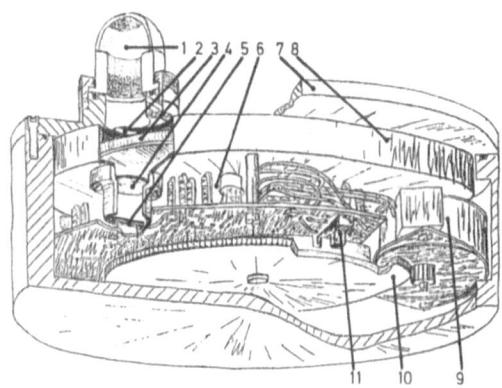

Fig. 2. Optical unit of quanta spectrometer.
1 = scattering dome; 2 = entrance slit; 3 = circular variable filter; 4 = lens; 5 = silicon photodiode;
6 = signal processing electronics; 7 = outer case; 8 = monochromator case; 9 = stepping motor; 10 = transmission gear;
11 = sensor for monochromator position.

light the vertical attenuation differs considerably from the attenuation measured in a beam of parallel light. The vertical attenuation also shows a strong dependence on solar altitude and degree of cloudiness (Sauberer, 1941).

The scattering dome described above has been used to develop a spectrometer with 2π response to incident light and a variable filter for spectral resolution. A cross-section of the optical unit of the spectrometer is depicted in Fig. 2. A prototype of this instrument was demonstrated at the Hannover Fair 1978, it will be commercially available in 1980.

Mathematical description of the relation between light and photosynthesis

A relation between photosynthesis and light intensity can be obtained from depth profiles of productivity and accompanying light measurements if there exists no time dependence of light inhibition. Harris & Piccinin (1977) found that light inhibition in confined samples can be much higher than in natural populations, where individual algae travel through a strong gradient of light within short periods. Our own experiments with samples from Steinhuder Meer did not indicate any decrease of light inhibition from normal exposure times down to 5 minutes, so it seems that in our case the mathematical formulation shown can be applied to experimental data obtained from depth profiles.

Vollenweider (1965) discusses some interpola-

tion equations describing the relation between light intensity and phytosynthesis. From a modification of these equations two different mathematical models will be derived in the following paragraph; their usefulness will be discussed using light data obtained from the recording station described above.

Equation (1) was first published by Smith (1936):

$$k \cdot I = P \cdot (P_{max}^2 - P^2)^{-1/2} \tag{1}$$

where P: photosynthetic rate per unit volume of water, and I: light intensity.

Equation (1) can be resolved for photosynthesis:

$$P = P_{max} \cdot \frac{k \cdot I}{1 + k^2 \cdot I^2} \tag{1a}$$

The equation describes a saturation effect; i.e. with increasing light intensity photosynthesis asymptotically reaches a maximum photosynthetic rate, P_{max}. Often a decrease in photosynthesis is measured at higher light intensities. Therefore Vollenweider (1965) introduced equation (2):

$$P = P_{max} \cdot \frac{k_1 \cdot I}{\sqrt{1 + k_1^2 \cdot I^2}} \cdot \frac{1}{(\sqrt{1 + k_2^2 \cdot I^2})^n} \tag{2}$$

Using this equation, a great variety of graphs can be represented by variation of the four constants P_{max}, k_1, k_2, and n (where n is an integer). But it is questionable whether an equation with so many coefficients can be used for interpolation of experimental results. We found a 2σ error for our

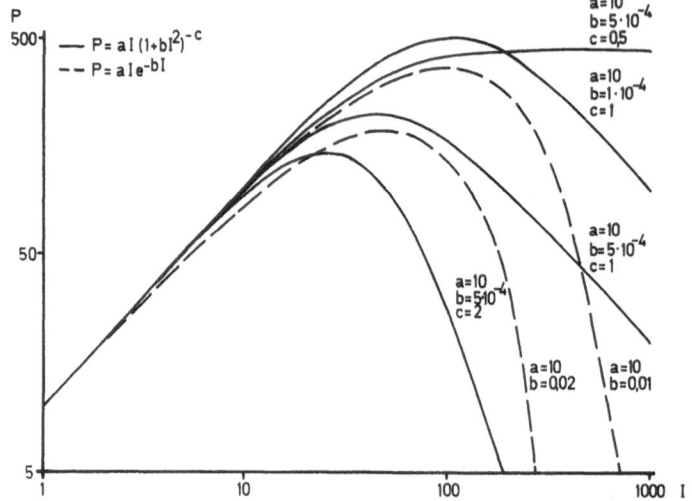

Fig. 3. Variation of coefficients a, b, c for equations (3) and (4). Light intensity I and photosynthesis P in arbitrary units.

C-14 photosynthesis measurements of about 20%. Therefore any interpolation formula which gives an approximation to the experimental results of 20% or better can be used as well.

Equation (2), which allows more "fine tuning" than necessary for practical application, can be simplified. With $k_1 = k_2$, and with renaming the constants, one gets:

$$P = a \cdot I \cdot (1 + b \cdot I^2)^{-c} \qquad (3)$$

There is no restriction on c, so c can be any real number $\geq \frac{1}{2}$. With $c = \frac{1}{2}$ equation (3) is identical to equation (1a). For $c = 2$ one can use equation (4), which gives a similar graph and has some

mathematical advantages:

$$P = a \cdot I \cdot e^{-b \cdot I} \qquad (4)$$

From this equation maximum photosynthesis, P_{max}, and corresponding optimum light intensity, I_{opt}, are easily calculated:

$$P_{max} = \frac{a}{b} \cdot \frac{1}{e} \qquad (5)$$

$$I_{opt} = \frac{1}{b} \qquad (6)$$

Figure 3 shows the influence of variation of coefficients a, b, and c on the graph described by equations (3) and (4). Variation of a controls the

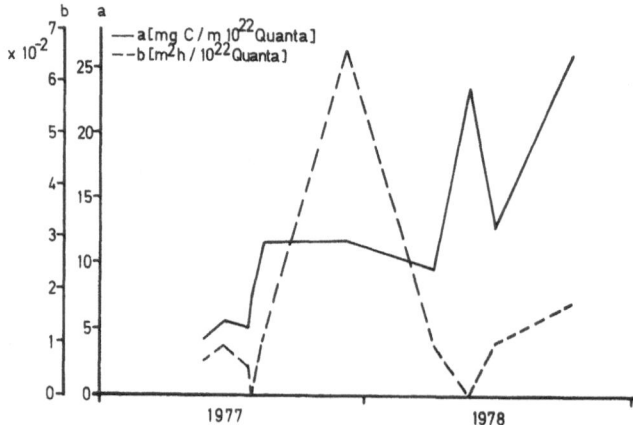

Fig. 4. Interpolation of primary productivity measurements using equation (4). Annual variation of coefficients a and b.

22

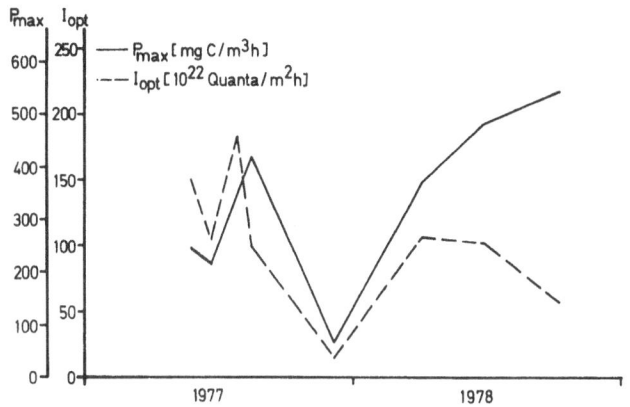

Fig. 5. Annual variation of P_{max} and I_{opt} calculated from equations (5) and (6).

slope of the photosynthesis light curve, b influences the maximum photosynthesis. For $b = 0$ no maximum is found, photosynthesis steadily increases with increasing light intensity. For $b \neq 0$ the amount of inhibition at higher light intensities is fixed in equation (4), whereas it varies with c in equation (3).

Figure 4 shows the annual variation of coefficients a and b, calculated from equation (4) for primary productivity measurements in Steinhuder Meer. There are two cases with $b = 0$, where light intensity was insufficient to reach saturation even in 10 cm depth. Figure 5 shows the variation of P_{max} and I_{opt} calculated from equations (5) and (6). In both years the maximum photosynthesis increased during the year. In 1978 it reached much higher values than in 1977. This can be attributed to the fact that the 1978 curve includes a value from October, whereas in 1977 the August value is directly followed by the December value. The maximum productivity most probably increased after August and reached similar values as in 1978. The I_{opt}-values for 1978 are slightly lower than the values for 1977. A reason for this could be the cloudy weather of 1978, so the algae – or the species composition – might have been adapted to lower light intensities.

Calculation of primary productivity from continuous light measurements between C-14 experiments

A comparison between measurements of depth profiles in Steinhuder Meer at midday and during the afternoon of the same day showed that results matched the same photosynthesis light curve. Thus the diurnal sum of production can be calculated from a continuous record of light intensity, using a photosynthesis light curve obtained at midday. For this purpose the coefficients for interpolation equation (3) or (4) must be calculated from the data obtained from the C-14 productivity measurement. The calculation was done by an approximation method using a small computer. Correlation coefficients usually were above 0.95, where equation (3) yielded a slightly better correlation than equation (4).

As mentioned above, integrated light intensity was recorded in 2 hour intervals. Therefore production per square meter was calculated in 2 hour intervals and summed up for the whole day. For calculation of primary production under a unit surface area the production light curve must be converted into a production depth curve, using the light intensity and transmission data for the respective 2 hour interval. Then the production depth curve must be integrated. This can be done in an elementary way for equation (4); equation (3) can only be integrated by using numerical methods, which again implies the use of a small computer.

The question arises whether the mathematical formulations shown can be used for interpolation of primary productivity measurements in intervals of several weeks. Figure 6 shows the results of C-14 measurements during 3 weeks in the period from 27 July 1977 to 15 August 1977. In this period the slope of the photosynthesis light curve

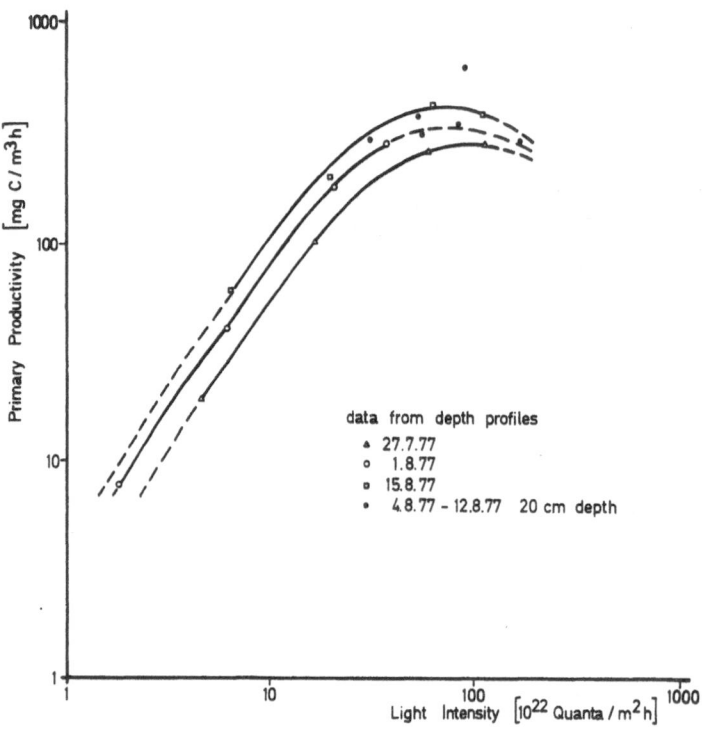

Fig. 6. Photosynthesis light curves obtained from depth profiles measured during a 3 week interval.

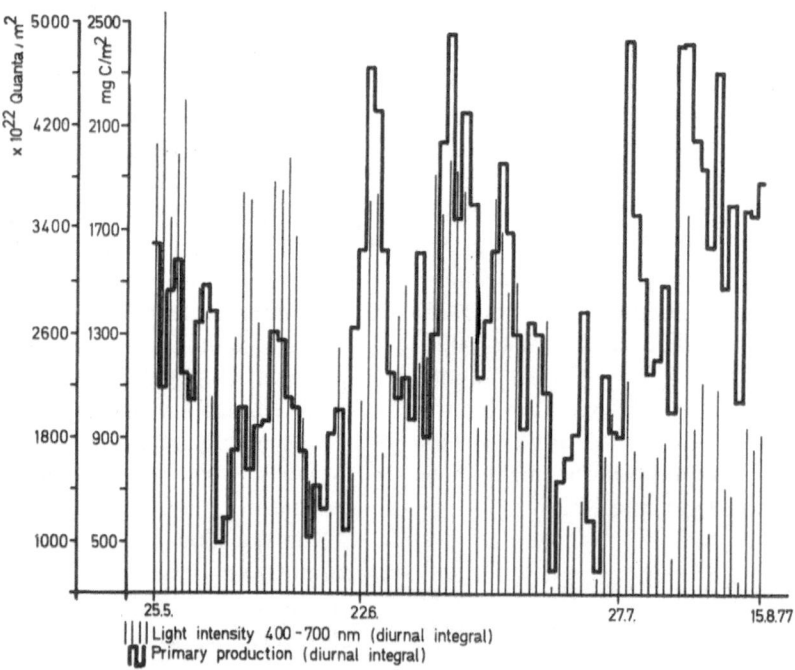

Fig. 7. Interpolation between productivity measurements using equation (4). Diurnal light integral (summed up from 2 hour intervals) and corresponding primary production.

24

and the maximum photosynthesis increased steadily.

During the period from 1 August to 15 August several measurements were made at 20 cm depth. Within the limits of accuracy all results fall between the two curves from beginning and end of this period. This leads to the conclusion that an interpolation between productivity measurements on the Steinhuder Meer in intervals of several weeks should be possible. Changes of other factors influencing primary productivity are slow compared to the variations of light intensity. The relative slow changes of P_{max} values (Fig. 5) lead to the same conclusion.

Primary production was calculated for the intervals from 24 May 1977 to 15 August 1977 and 5 April 1978 to 10 August 1978 using both equations (3) and (4). In Fig. 7 variation of diurnal production calculated from equation (4) is compared with total underwater irradiance in quantum units. There are pronounced variations in production, which do not always follow the variations of light intensity. This is due to the nonlinear relation between light and productivity. Especially the duration of light inhibition varies from day to day with varying degree of cloudiness.

The total production from 25 May to 15 August 1977 calculated from equation (3) differs only 0.1% from the value calculated from equation (4). Simple linear interpolation between the C-14 productivity measurements on the indicated four days gives a value which is 7% higher. This seems to be an accidental good correspondence. From April to August 1978 the results obtained by using both equations again differed by only 2%, whereas linear interpolation between measurements gave a value 32.5% lower.

Conclusions

The mathematical formulations shown make it possible to calculate the diurnal sum of production by using a continuous underwater light record and measuring a depth profile of primary productivity at midday. Furthermore, production can be calculated for the interval between productivity measurements. Both equations (3) and (4) gave the same results for total production. Differences between the results obtained by calculation from light measurements and simple linear interpolation between monthly productivity measurements can be as high as 30%. This has to be confirmed by further measurements.

Acknowledgements

The author wishes to thank Dr. H. W. Seibold, who took the trouble to convert the light data from the chart recorder into a computer readable form, and Mrs. S. Kretschmer, who typed part of the light data and made the illustrations.

References

Bjoern, L. O. 1975. Photobiologie, Gustav Fischer Verlag, Stuttgart. 210 pp.

Booth, C. R. 1976. The Design and Evaluation of a Measurement System for Photosynthetically Active Quantum Scalar Irradiance. Limnology and Oceanography 21 (2): 326–335.

Gessner, F. 1959. Hydrobotanik Bd. II. Deutscher Verlag der Wissenschaften, Berlin. 701 pp.

Golterman, H. L. & Clymo, R. S. 1969. Methods for chemical Analysis of Fresh Waters. IBP Handbook No. 8. Blackwell Scientific Publications, Oxford. 172 pp.

Harris, G. P. & Piccinin, B. B. 1977. Photosynthesis by Natural Phytoplankton Populations. Arch. Hydrobiol. 80 (4): 405–457.

Heath, O. V. S. 1972. Physiologie der Photosynthese. Georg Thieme Verlag, Stuttgart. 314 pp.

Levin, G. V. & Lindgren, W. A. (dates not given). Automated Primary Productivity Instrument. Biospherics Incorporated, Rockville, U.S.A.

Lewis, W. M. 1975. A Theoretical Comparison of the Attenuation of Light Energy and Quanta in Waters of Divergent Optical Properties. Arch. Hydrobiol. 75 (3): 285–296.

Sauberer, F. & Ruttner, F. 1941. Die Strahlungsverhältnisse der Binnengewässer. Akademische Verlagsanstalt, Leipzig. 240 pp.

Sauberer, F. & Härtel, O. 1959. Pflanze und Strahlung. Akademische Verlagsgesellschaft Leipzig. 268 pp.

Sauberer, F. 1962. Empfehlungen für die Durchführung von Strahlungsmessungen an und in Gewässern. Mitt. internat. Verein. Limnol. No. 11, Stuttgart.

Smith, E. L. 1936. Photosynthesis in relation to light and carbon dioxide. Proc. Nat. Acad. Science, Wash., 22: 504.

Vollenweider, R. A. 1965. Calculation Models of Photosynthesis-Depth Curves. Mem. Inst. Ital. Idrobiol., 18: 425–457.

Vollenweider, R. A. 1969. Primary Production in Aquatic Environments. IBP Handbook No. 12. Blackwell Scientific Publications, Oxford. 213 pp.

Wallen, D. G. & Geen, G. H. 1968. Loss of Radioactivity During Storage of C-14-Labelled Phytoplankton on Membrane Filters. J. Fish. Res. Bd. Canada, 25 (10): 2219–2224.

Weinberg, S. 1974. A Relative Irradiance Meter for Submarine Ecological Measurements. Netherlands Journal of Sea Research 8 (4): 354–360.

FLUOROMETRIC CHLOROPHYLL ESTIMATION OF VARIOUS ALGAL POPULATIONS

B. GEORGI, E. SCHULZE & D. ERNST

Abstract

Fluorometric determination of the chlorophyll content in natural lake populations can be determined by a new fluorometer. Its construction fulfills the requirements of limnological field experiments for high sensitivity, easy portability and ability to measure various photosynthetic parameters. Measurements are performed with the help of a special excitation mode and photosynthetic blocking with CMU. The measurements indicate the primary production of natural algal populations.

Introduction

The in situ fluorometric determination of chlorophyll in phytoplankton populations has been used in laboratory as well as in field experiments (Vollenweider, 1969; Lorenzen, 1966). Measuring difficulties like low concentration ranges and non-portable equipment have limited a general introduction into field work (Vollenweider, 1969). During the studies on primary production of the Steinhuder Meer with ^{14}C (Steemann Nielsen, 1952) since 1972, we tried to apply our experiences in fluorometric analyses in these investigations. The construction of the portable and highly sensitive fluorometer (Schulze, 1978) includes the special interdependence of photosynthesis and fluorescence. (Papageorgiou, Gowindjee, 1969). Using flash-excitation after a fixed dark period, the Kautsky effect can be measured. In combination with the blocking effect of

CMU on photosynthesis, the fluorometric signals correlate with the chlorophyll concentration and the physiological conditions of the algae. Results with algal cultures and lake water from the Steinhuder Meer demonstrate the practical applicability of the fluorometer.

Material and methods

Algal cultures of *Chlorella vulgaris, Chlorella fusca, Scenedesmus quadricauda, Diatomeaen nitzschia commutata, Anabena variablis* and *Nostoc muscarama* were obtained from the algal collection of the Institute of Plant Physiology, University of Göttingen and the Institute of Botany, Tierärztliche Hochschule Hannover.

The ^{14}C fixation measurements were performed by a modified Steemann Nielsen method. The chlorophyll determination was according to Arnon (cit. Urbach, 1976). For the cell counting we used a THOMA chamber.

The fluorescence spectrum and the first measurements were recorded with the BAIRD-ATOMIC FLUORISPEC SF 100. The special fluorometer was manufactured in the workshop of the institute. The field experiments were carried out on the Steinhuder Meer, west of Hannover.

The fluorescence spectrum of a sample from the Steinhuder Meer was recorded (Fig. 1). A high resolution fluorometer requires filters with narrow bandwiths and high transmission values. Single filters for such purposes are not available. There-

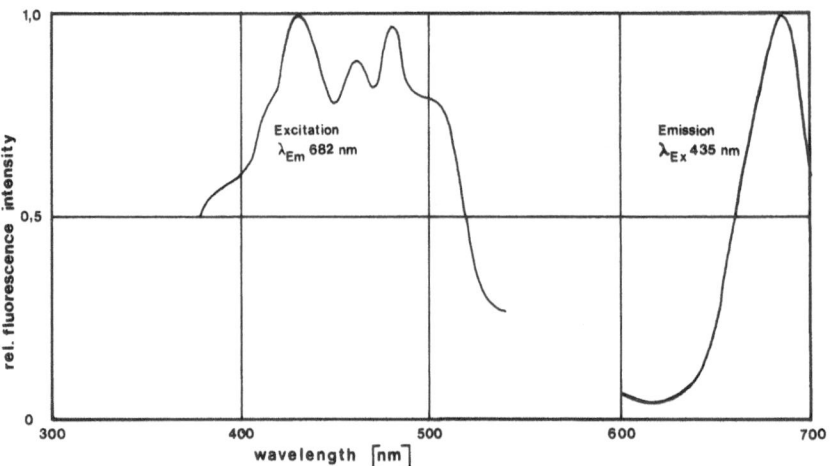

Fig. 1. Fluorescence spectrum of a sample from the Steinhuder Meer (corrected).

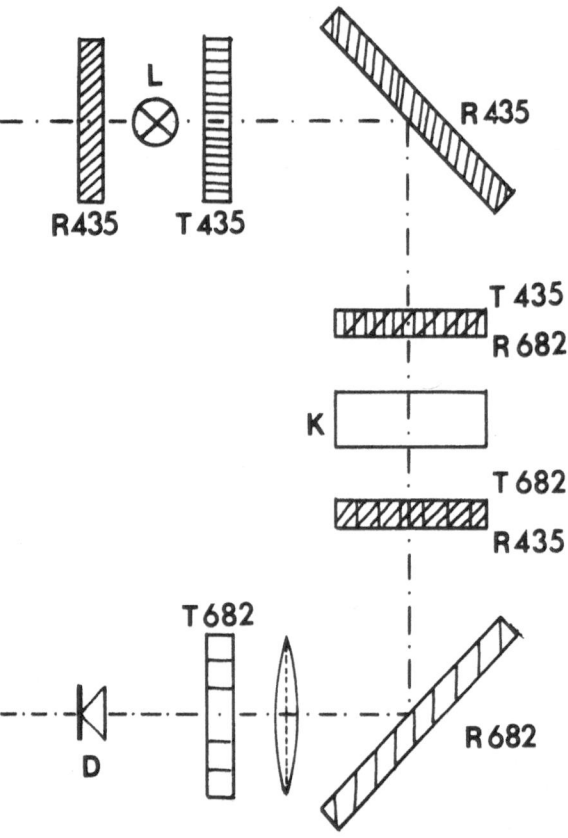

Fig. 2. Light pass of the fluorometer. The excitating light from the lamp (L) is selectively reflected by the mirrors (R 435) and filtered by T 435. It induces fluorescence light in the cuvette (K). This light is selectively reflected by the mirrors (R 682), filtered by special glasses T 682 and focused by a convex lense on to the photodiode (D).

fore, a combination of commercial interference filters (AL 434 and AL 677, Fa. SCHOTT) and specially manufactured selective mirror-filters (R 435 and R 682) manufactured by Institut für Angewandte Physik, Universität Hannover, corresponding to the spectrum of natural algal populations were selected. The principle of the light pass is shown and described on Fig. 2.

The optical arrangement guarantees a measurement of natural algal populations without any concentration steps. The rather big sample of 25 ml in a cuvette with a diameter of 40 mm is focused on the detector to avoid sampling errors by inhomogenous distribution and sedimentation (Christoffers, 1976). The high sensitivity of photodiodes connected with high resolution amplifiers is comparable to that of photomultipliers (Georgi, 1977) and does not require a stabilized high voltage power supply. According to the special construction, the fluorometer can be used even on small boats (Fig. 3).

Results

The practical measurements started with cultures of *Chlorella vulgaris* and *Scenedesmus*. The calibration curves for the number of algae (Fig. 4a) and the chlorophyll$_{a+b}$ (Fig. 4b) concentration correlate with the fluorescence over a fairly wide range.

The linear range covers the cell numbers or chlorophyll contents found in the lakes

Fig. 3. Fluorometer in operation on a small boat.

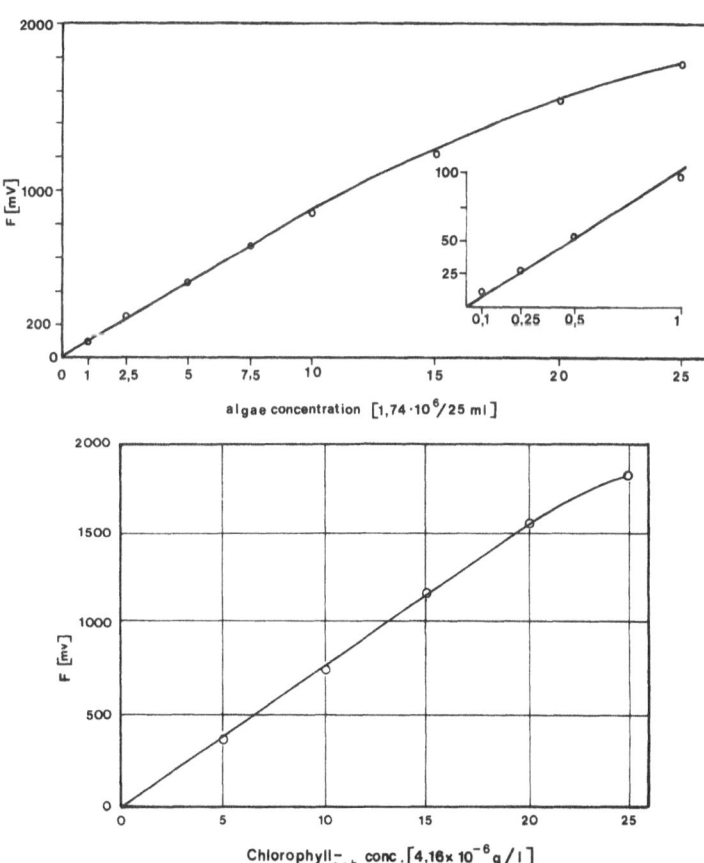

Fig. 4. Correlation between number of algae and fluorescence signal (a) of a 3-day old algal population (Chlorella vulgaris). Correlation between chlorophyll$_{a+b}$ concentration and fluorescence signal (b).

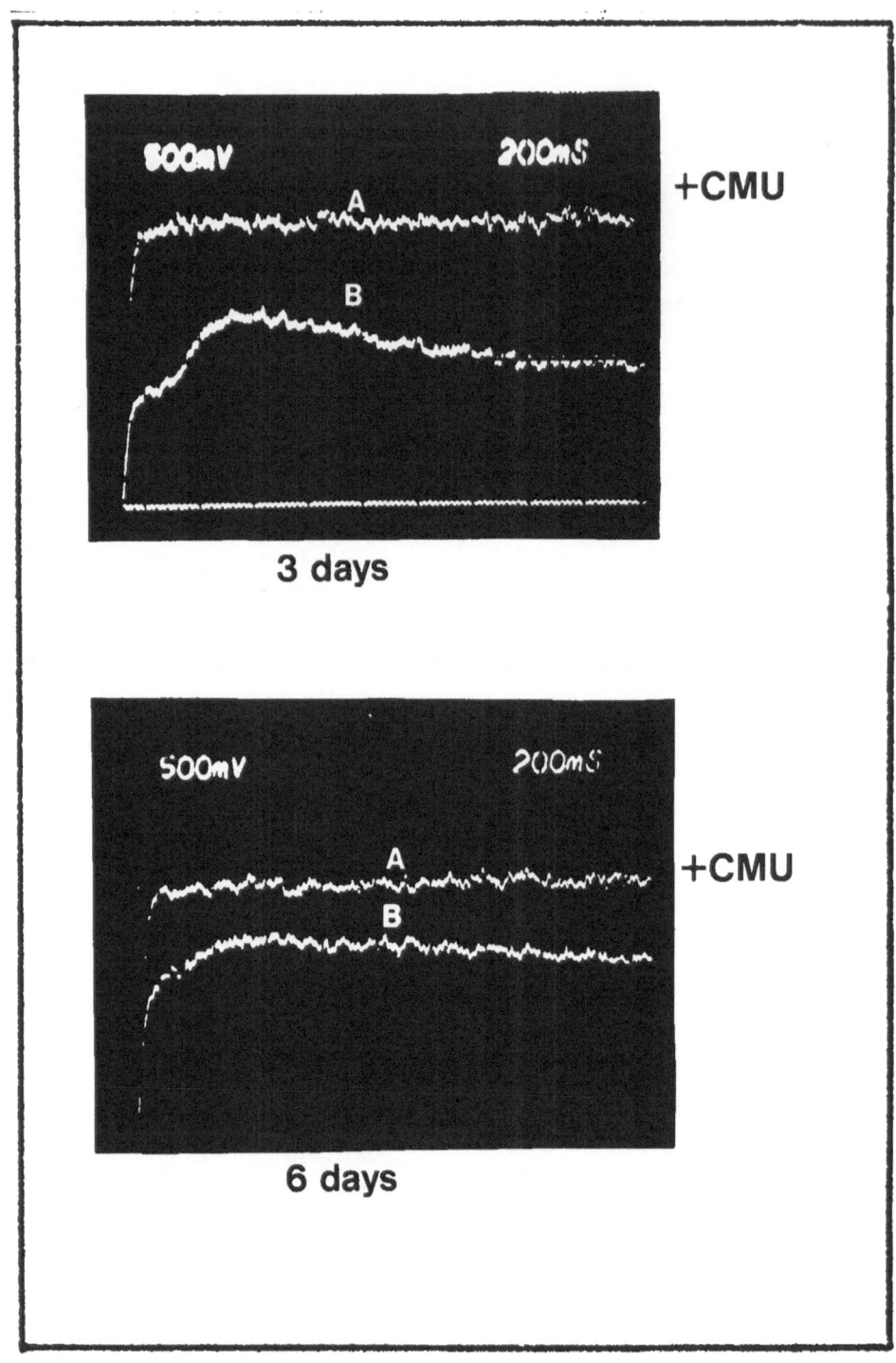

Fig. 5. Time course of the fluoroscence signals of algal populations of different age (Chlorella vulgaris)
Curves A recorded after poisoning with CMU
Curves B show the courses of health algae.

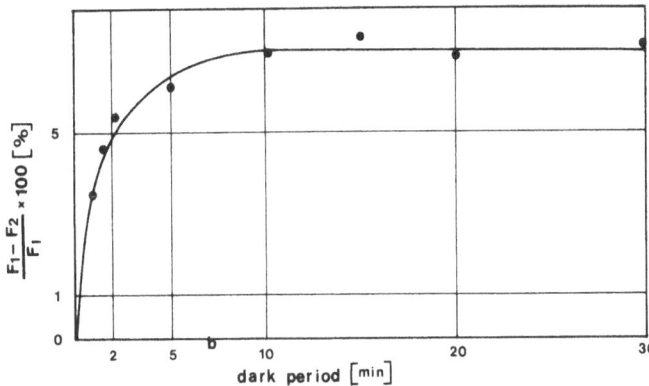

Fig. 6. Standardized differences of the fluoroescence signals induced by the first (F_1) and
the following flash (F_2) after various dark periods.

(Steinhuder Meer and Dümmer See) with a typical value around F = 200 mV.

The reproducibility of the fluorescence measurements is better than ±1.25% after calibration with an external glass standard. The standard (OG 550, Fa. Schott) had a similar temperature gradient as the chlorophyll pigment.

The quantum yield of fluorescent living cells decreases with increasing photosynthetic activity demonstrated by the time course of the fluorescence signal (Fig. 5). After application of CMU the fluorescence signal is independent of the physiological conditions of the algae and is solely related to the chlorophyll concentration. The comparison of the increased fluorescence and the primary production shows comparable results for algal cultures as well as natural lake populations. The results are in good agreement with those published by Samuelsson *et al.*, 1978. Additionally, we found that it is necessary to keep the sample in darkness for a certain period to obtain the right value for the unpoisoned algae. Excitations by single flashes of 3 msec at intervals of 10 sec induce differences between the fluorescence values of the first and the following excitations (Fig. 6).

The maximum of the difference will be attained after a dark period of 10 min before the first excitation. This result, which is based on the time course of the fluorescence yield immediately after excitation, has been found by Kautsky (1943). A correlation between the amount of this effect and the age of an algal population is shown in Fig. 5 (curves B). A practical advantage will be achieved

by comparing the fluorescence differences induced by poisoning with CMU and the described excitation mode shown in Fig. 7 for a growing culture of *Chlorella fusca*.

Testing natural populations these methods correlate with a factor of 0.9 when the measurements are performed immediately after sampling on the lake.

Combining these results we estimated the chlorophyll content in the Steinhuder Meer, on the basis of a *Scenedesmus quadricauda* culture with the same fluorometric behaviour, to have an average of 0.98 g m^{-3} for May 1979.

Discussion

The correlation factor between the fluorescence values for CMU poisoned samples and for the described excitation mode was reduced if the measurements were taken after transportation, even though a temperature controlled dark box was used. The easy handling of the instrument allows routine chlorophyll determinations to be carried out on cultures or lake populations after calibration with the traditional chlorophyll estimation method. The sensitivity provides a chlorophyll determination under field conditions without concentration steps. Other possibilities are the detection of substances which interfere with the photosynthetic process. This is done by addition of healthy algal populations with known KAUTSKY- and CMU-behaviour. The studies with decoupling agents like methylamin will provide further information on the photosynthetic capacity.

31

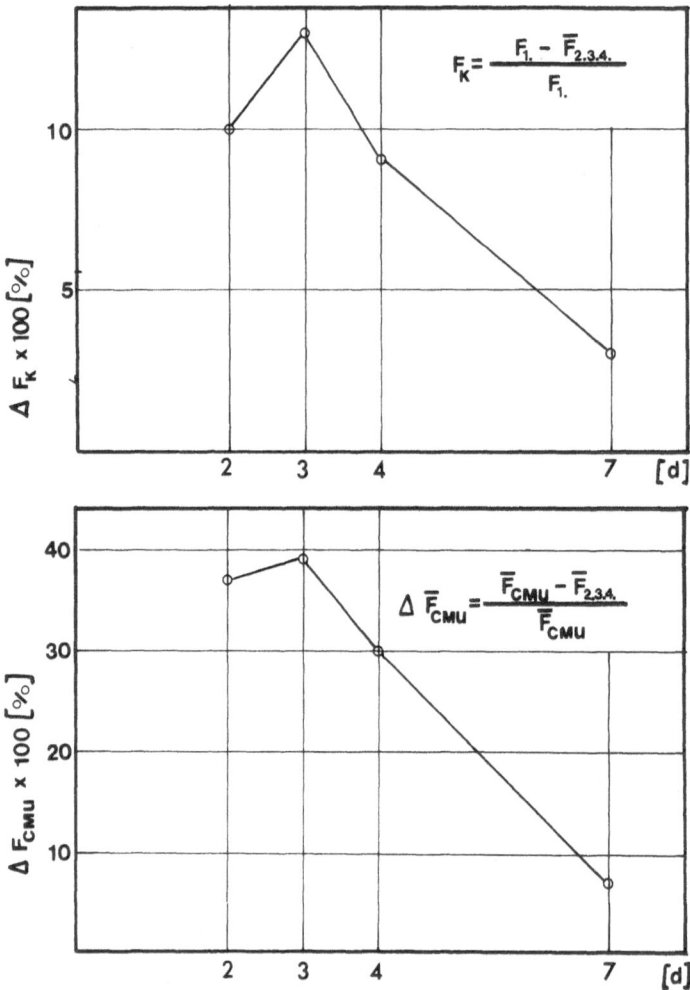

Fig. 7. Standardized differences between the first flash-induced fluorescence value (F_1) and the average of the following ($\bar{F}_{2,3,4}$) for different ages of a Chlorella fusca culture (a) and the standardized differences between CMU-poisoned values ($\bar{F}_{CMU}$) and $\bar{F}_{2,3,4}$ (b).

First comparisons between primary production measurements with the ^{14}C method and the photosynthetic capacity determined by in vivo fluorescence under field conditions are promising, but need further verification in long-term experiments.

References

Christoffers, D. 1976. Fluorometrische Lysinbestimmungen in Getreidemehlen. Thesis TU Hannover. 12–16.

Georgi, B. 1977. Fluorometrische Bestimmung von Aminosäuren. Thesis TU Hannover. 40–43.

Kautsky, H. & Frank, U. 1943. Chlorophyllfluoreszenz und Kohlensäureassimilation. Biochem. Z. 315: 139–232.

Lorenzen, C. J. 1966. A method for the continuous measurement of in vivo chlorophyll concentration. Deep-Sea Res. 13: 223–227.

Papageorgiou, G. & Govindjee, I. 1969. The second wave of fluorescence induction in Chlorella pyrenoidosa. Progress in Photosynthesis Research II: 905–912.

Samuelson, G., Öquist, G. & Halldal, P. R. 1978. The variable chlorophyll(a) fluorescence as a measure of photosynthetic capacity in algae. Mitt. Internat. Verein Limnol. 21: 207–215.

Schulze, E. 1978. Konstruktion und Erprobung eines für in vivo Messungen von Chlorophyll geeigneten Impulsfluorometers. Diplomarbeit TU Hannover. 1–57.

Steemann Nielsen, E. 1952. The use of radioactive carbon (^{14}C) for measuring organic production in the sea. J. Cons. Int. Explor. Mer. 18: 117–140.

Urbach, W., Rupp W. & Sturm. H. 1976 Experimente zur Stoffwechselphysiologie der Pflanzen. G. Thieme Verlag. 96–98.

Vollenweider, R. (ed), 1974. A manual on methods for measuring primary production in aquatic environments Blackwell Scientific Publ. 14–31.

2. NEUSIEDLERSEE, AUSTRIA

BACKGROUND DATA

Latitude: 47°38'–47°51'N
Longitude:16°41'–16°52'E
Altitude: 113 m a. S.L.
1 (km) 35
b (km) 12
L (km)
L_D
Name of the main tributary: Wulka
Average inflow m³/sec.
Average outflow m³/sec.
Theoretical retention time

origin: tectonic
z (m) 1.8
$\bar{z}$ (m) 1.3
$\bar{z}/z$
V (km³) 200–250
A (km³) 300
A' (km²)
A' : A

Geological characteristics

Crystalline basement, quaternary sediments and gravels

Climatic conditions

Average monthly temperature 1.5, 0.4, 4.4, 10.5, 14.6, 18.5, 20.0, 19.4, 15.9, 10.6, 5.3, 0.7
Average precipitation/year 650 mm
Average sunshine duration
Main wind direction(s) NW, SE
Evaporation per year 900 mm

Ice cover (days) 54
Average radiation/year 1×10^6 Kcal m^{-2}
% of calm days

Cultural geography and demography

Land usage of catchment area (%)
Industrial
Agricultural 32 + 34 vineyards
Meadows 21

Usage of lake water including recreational activities
Fishery; Boating; Sailing; Swimming

Forest 7.8
unused 5.2
number of residents   inhabitants/km^2

Water temperature:	min 4	max 32
Secchi depths:	min 0.08 m	max 0.52 m
Euphotic zone:	min 0.37 m	max 1.80 m
O$_2$ concentration: % saturation	min $\varnothing$	max 105

pH 7.5–10   Conductivity (μS) 1000–2300   Alkalinity (mval) 8–10.5

Average P-conc. 0.14 mg $\cdot$ l^{-1} Average N-conc. 1.3–2.1 mg $\cdot$ l^{-1}

Conditions of sediment:

Soft mud (more than 60% water content) in the northern and western part of the lake. Compact mud and sand on the eastern shore. Oxic to anoxic; especially in the reed belt.

Dom. phytoplankton species:

Monoraphidium contortum, Dictyosphaerium ehrenbergianum, Oocystis lacustris, and various other green algae, Cyclotella meneghiniana, Cryptophytes, recently Microcystis pulvera

Dom. zooplankton species:

Arctodiaptomus spinosus, Diaphanosoma brachyurum, Brachionus angularis, Keratella quadrata, K. cochlearis, Synchaeta tremula-oblonga group and Hexarthra fennica, recently Rhinoglena fertöensis

Dom. macrophytes:

Phragmites communis,
Potamogeton pectinatus, Myriophyllum spicatum

Dom. benthic organisms:

Nematoda, Tardigrada, Chironomida, Rotatoria

Fishes:

Anguilla anguilla, Stizostedion lucioperca, Esox lucius, Cyprinus carpio, Ctenopharyngodon idella, Hypophthalmichthys molitrix, Pelecus cultratus, Blicca björkna, Acerina cernua

PHOSPHORUS AND NITROGEN IN NEUSIEDLERSEE

F. NEUHUBER, P. ZAHRADNIK & H. BROSSMANN

Abstract

In the last few years the nutrient content of Neusiedlersee has increased considerably. This process has a stepwise character starting with an increase of dissolved nutrients followed by an increase in the biomass, but it does not include the sediment because of the erosion of the surface sediment layer and its transport into the reed belt. The influence of two nutrient sources (rain, inflow) is reported.

Introduction

Investigations of phosphorus and nitrogen in Neusiedlersee show a big spatial and temporal variation in their concentrations as a result of changes in the physico-chemical conditions, biological conditions, and nutrient loadings of the incoming water.

Results and discussions

Physico-chemical aspects of Neusiedlersee

Physico-chemical conditions that influence the nutrient content of Neusiedlersee include the high content of suspended matter, the loss of orthophosphate by an inorganic process, and the deposition of suspended matter in the reed belt.

A. The high content of suspended matter

The influence of wind leads to a change of suspended matter. The suspended solids vary from 10

to 500 mg l^{-1} and more. Parallel to this variation the particulate phosphorus varies in the same manner (Fig. 1).

Deviations that are detected when either comparing different sampling points within a series or comparing different series, depend on the composition of the suspended matter. The material of the sediment surface has a much lower phosphorus content than planktonic organisms. In Neusiedlersee sediment phosphorus constitutes 0.2–0.6% of the dry weight (d.w.), while organisms have a phosphorus content of 1% d.w. (Golterman, 1975). Since the lake water contains enough phosphate during the whole year, that an impoverishment of the phytoplankton by phosphate can be excluded, a transgression of the phosphorus value of the sediment must be taken into account with phosphorus derived from planktonic organisms. On the whole the particulate phosphorus expressed in parts per thousand of d.w. is a criterion of an increase in the biomass, which we call eutrophication, though it should be pointed out that the ratio of planktonic phosphorus to phosphorus derived from the sediment depends mainly on the wind situation.

B. The loss of orthophosphate by an inorganic process

The organic content of the sediment is very low (Table 1). From analysis of amino acids about one quarter of the Kjeldahl-nitrogen is in the form of aminoacids. Therefore the content of living

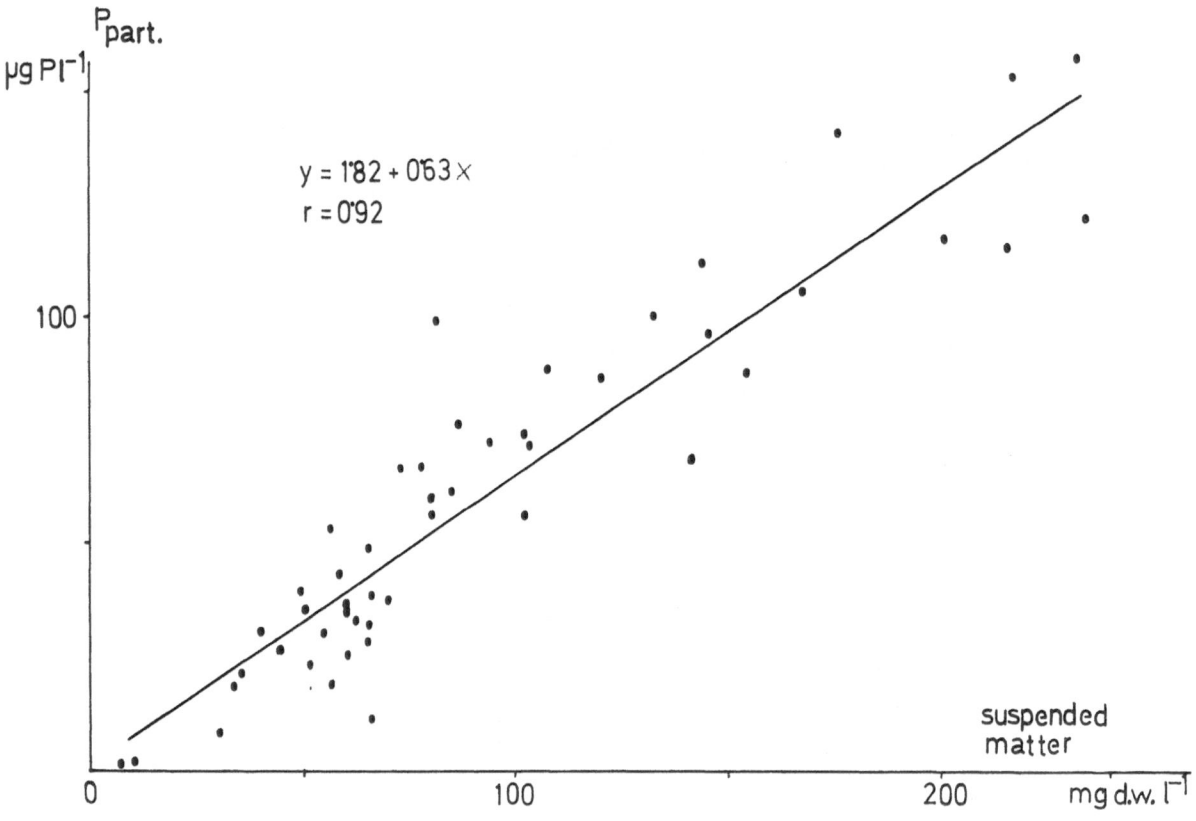

Fig. 1. The relation between suspended matter and particulate phosphorus of three sample series.

material in the sediment is very low. One tenth of carbon, two thirds of nitrogen and one eighth of phosphorus can be attributed to living matter, if it is assumed from the amino acids that 0.7% d.w. is living material.

The greater part of sediment carbon must be humic matter with a low nitrogen and phosphorus content, a result of the complete mineralization in this lake. The greater part of the phosphorus must occur in an inorganic compound. Experiments carried out with ^{32}P show a phosphate decrease in the water (supernatant) with increasing temperature (Fig. 2). Therefore we can conclude the occurrence of an inorganic process. The chemical conditions of the lake with a high pH (8–9) and the

Table 1. Carbon, nitrogen and phosphorus content of the lake sediment.

C:	2.2 –5.0 % d.w.*
N:	0.12–2.00 "
P:	0.02–0.06 "

* d.w. = dry weight

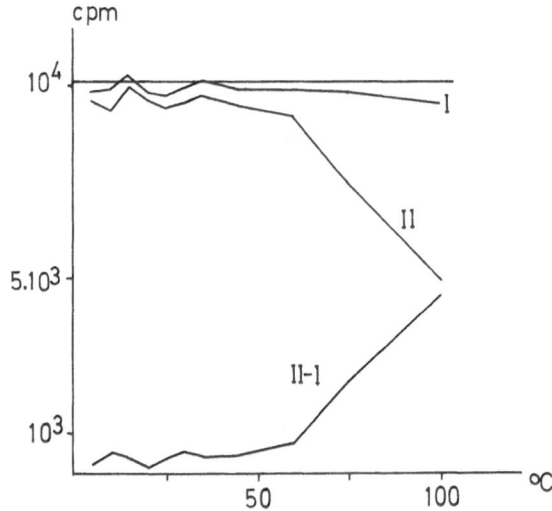

Fig. 2. Temperature dependence of $^{32}PO_4^{3-}$ decrease in the supernatant with a content of suspended matter of 350 mg d.w.l^{-1}. I Centrifugation of the sample before ^{32}P addition. II Centrifugation of the sample after 17' ^{32}P addition.

considerable aerogenic and biogenic calcium carbonate precipitation confirm this assumption. Also the dependence of phosphate uptake on the amount of suspended matter indicates such a process (Gunatilaka, 1978). It is assumed that the phosphorus containing particles are very small and that an unknown amount can pass the filters used with the result that it is detected as part of the dissolved phosphorus fraction.

C. Deposition of suspended matter in the reed belt

The suspended matter of the open lake water is transported into the reed belt and once settled is not returned to the waterbody. Sedimentation has a maximum in the transitional zone between the open lake and the reed belt, and then decreases within the reeds as distance from the inner edge of the reed belt increases (Table 2). From assumptions of the phosphorus content of the lake sediment we can calculate the minimum quantities of phosphorus deposited.

Table 2. Average sedimentation in the reed belt near Neusiedl.

Distance from the open lake	Dry weight $g \cdot m^{-2} \cdot d^{-1}$	Least phosphorus deposited, $mgP \cdot m^{-2} \cdot d^{-1}$
0 m	440	176
50 m	170	102
100 m	60	36
150 m	30	18

The nutrient conditions of Neusiedlersee

Due to the big changes in particulate phosphorus, resulting from variations in windspeed, the concentrations of total phosphorus cannot be used as an indicator of an increase in nutrients. However the dissolved phosphorus fraction (geometrical means) shows a clear increase from 1972 to 1975, and although there was a decrease in 1976, this was followed by a further increase in 1978 (Fig. 3). The same change holds for nitrate (Fig. 4).

The horizontal distribution of the dissolved phosphorus compounds demonstrates the influence of point sources (Figs. 5–8). Due to the distribution of the inflows the northern part of the lake contains higher concentrations than the southern part. The particulate phosphorus expres-

sed in % d.w. shows a similar pattern, possibly depending on circulation patterns. No remarkable change in the particulate phosphorus content of the suspended matter can be detected from 1972 to 1975 (1972: 0.4–0.6%, 1975: 0.5–0.75% d.w.). But with the decrease of the dissolved phosphorus fraction in 1976 the particulate phosphorus increased by more than the double in the whole lake (1.4–1.5% d.w.). This increase continued in 1977 (lake centre: 1.5–2.0% d.w.).

In 1978 no change in the particulate phosphorus can be observed, while the dissolved fractions rose. This stepwise and alternating increase of the dissolved and particulate phosphorus is initiated by point sources, especially by the Wulka river in the northwest of the lake. From these point sources the phosphorus content increases southwards down the lake.

The point sources do not have much influence on the nitrate and ammonia content of the lake, but another factor may be that the uptake by the phytoplankton is sufficiently rapid to disguise the input from point sources. In the case of ammonia we can assume a loss to the air caused by the high pH and turbulence of the water.

The phosphorus and nitrogen input by superficial inflows and by precipitation

The superficial inflows to the Austrian part of the lake comprise the Wulka river with 60% of the inflowing water load, while the remaining 40% comes from smaller inflows, partly receiving effluents of sewage plants. The Wulka river has been well studied and the results are summarized in Table 3.

The water input from direct precipitation is about three times higher than that from the inflows. Since the rain water has a high inorganic nitrogen content, the nitrogen addition is mainly from this source depending on the quantity of rain, except during March and April (Fig. 9), where the high values of the latter may be the result of fertilizing. The same holds for the concentrations of the inorganic nitrogen compounds (Fig. 10) but in contrast to the nutrient loads from the precipitation, which means the higher the concentration the lower the quantity of precipitation, also with exception of the months March and April.

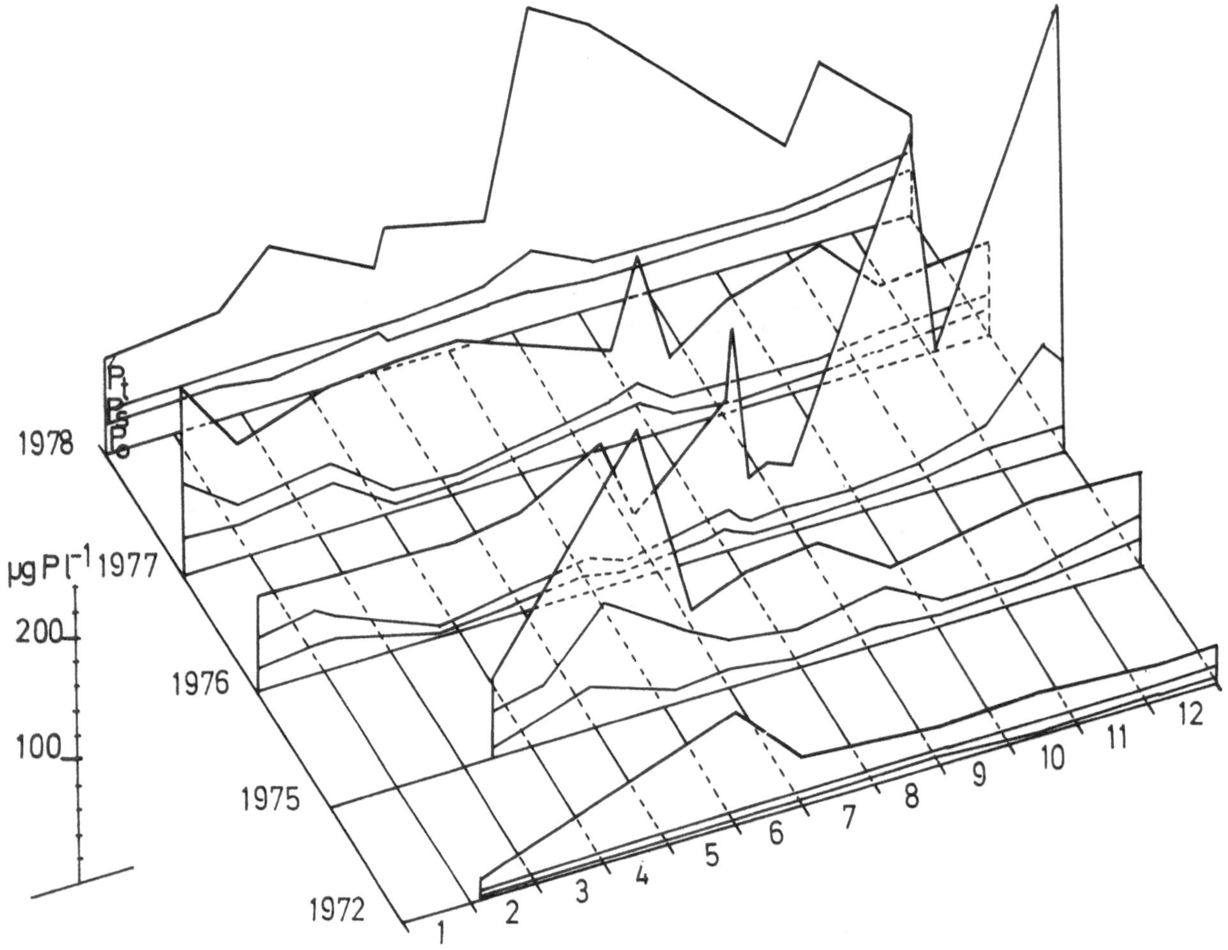

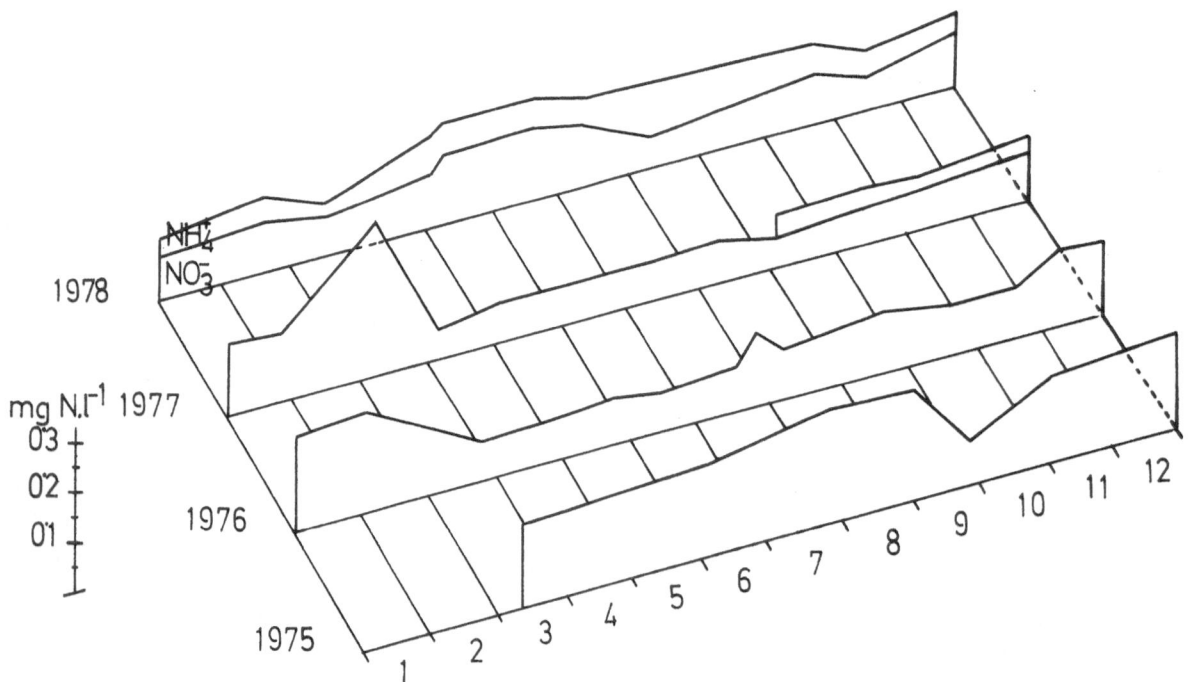

Fig. 3. Geometrical means of total phosphorus (P$_t$), dissolved phosphorus (P$_s$) and orthophosphate (P$_o$) of the sample series.

Fig. 4. Geometrical means of nitrate and ammonia of the sample series.

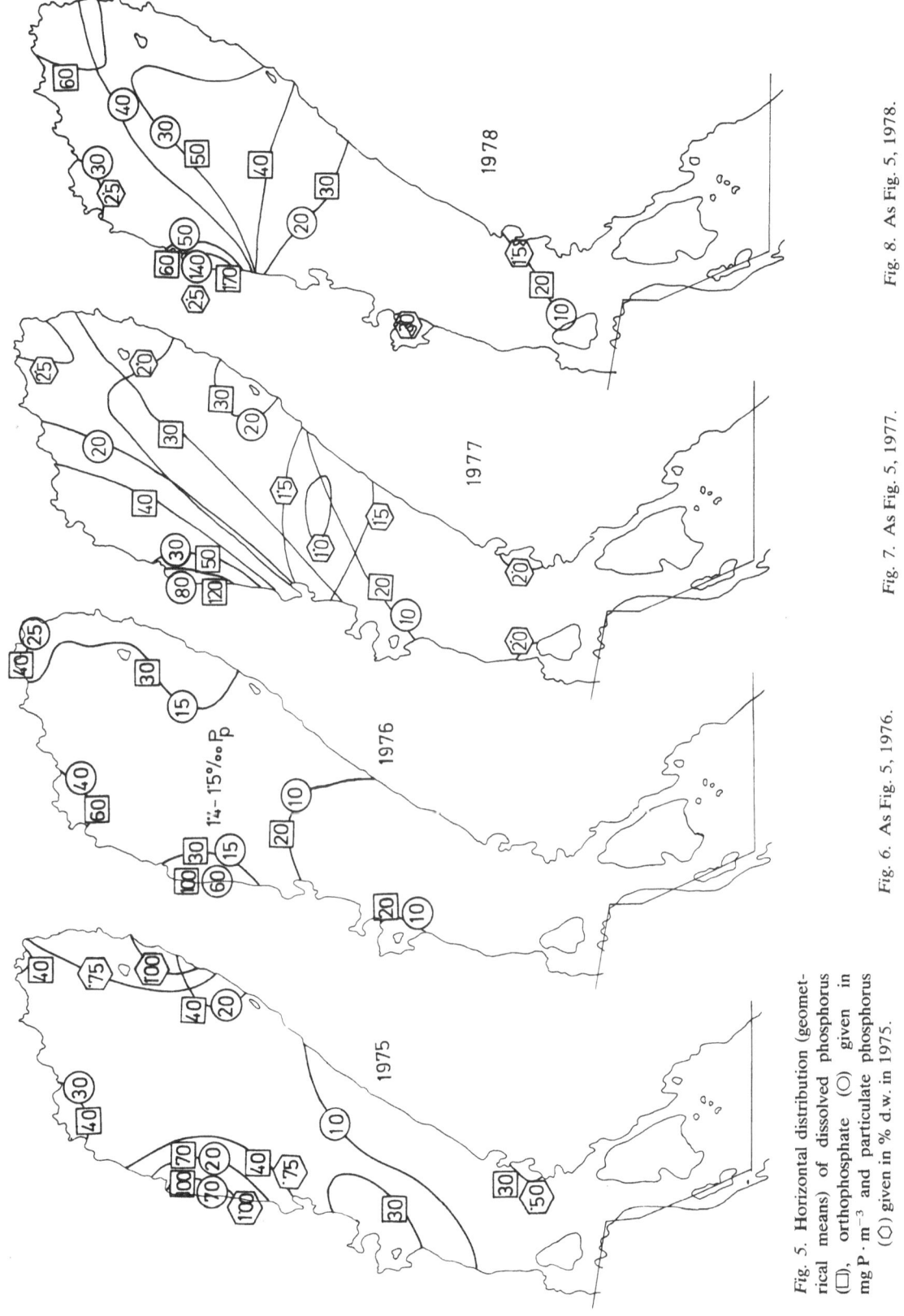

Fig. 5. Horizontal distribution (geometrical means) of dissolved phosphorus (□), orthophosphate (○) given in mg P · m⁻³ and particulate phosphorus (◯) given in % d.w. in 1975.

Fig. 6. As Fig. 5, 1976.

Fig. 7. As Fig. 5, 1977.

Fig. 8. As Fig. 5, 1978.

39

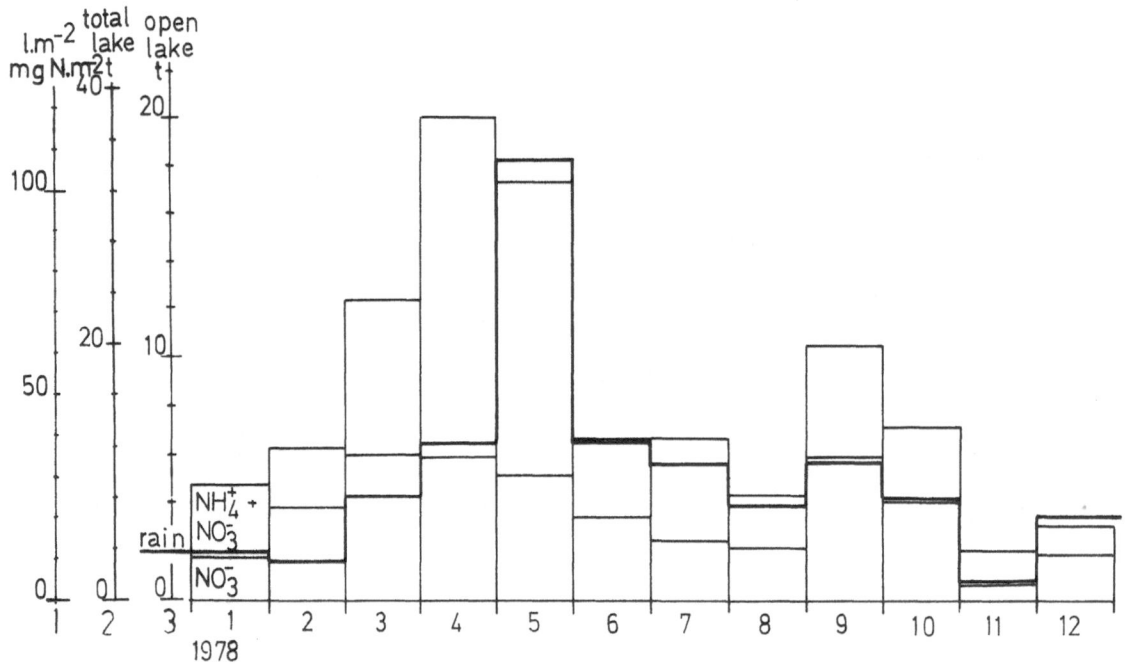

Fig. 9. Monthly input of water (thick line), nitrate and ammonia (thin lines) into Neusiedlersee by precipitation. Scale 1: precipitation in 1 per m² resp. nitrogen in precipitation in mg N per m², scale 2: Input of water resp. nitrogen to the total lake area given in tons, scale 3: Input of water resp. nitrogen to the open lake area given in tons.

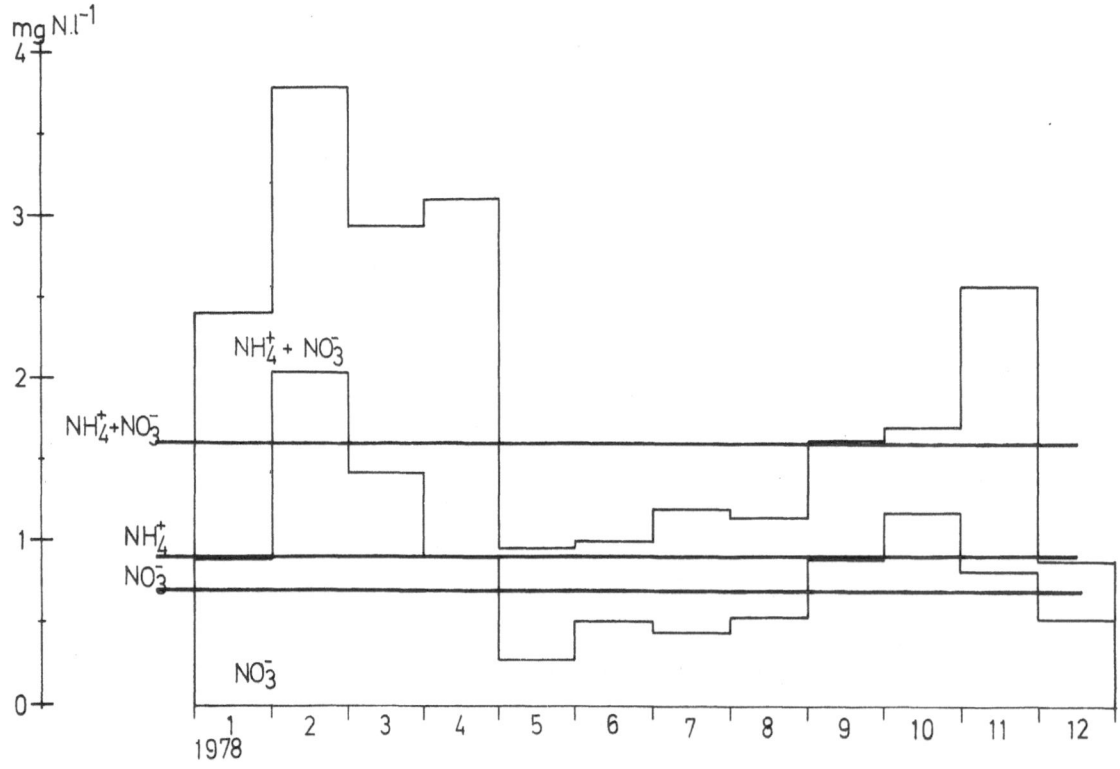

Fig. 10. Monthly mean concentrations and yearly mean concentrations of nitrate and ammonia of rain water.

Table 3. Mean nutrient content of the open lake in 1975 and 1978, yearly input by the Wulka river and by precipitation, yearly loss by Einser canal given in tons nitrogen resp. phosphorus rounded to 0.5 tons.

	Open lake 1975	Open lake 1978	Wulka	Rain total lake	Rain open lake	Einser c. 1975	Einser c. 1978
NH_4^+	—	8.5	40*	192	102	—	1.5
NO_3^-	30	17	35	132	71	5.5	3
P_{tot}	12	24	31	10	5.5	2	4
P_{diss}	5	7	20			1	1
PO_4^{3-}	2.5	3.5	12.5	2	1	0.5	0.5

* Evaluated from a smaller sample number

Though only 100 mm of the precipitation (one sixth of the yearly mean precipitation) was used for phosphorus analyses, the results obtained show a low phosphorus content of the rain water. The present results are summarized with the nitrogen data in Table 3. The changes of the nutrient content of the open lake calculated from the mean concentrations (Tables 3 and 4) show the decrease of nitrate and the increase of the different phosphorus fractions. Total phosphorus was calculated from the yearly mean of suspended solids ($50 \, mg \cdot l^{-1}$) and the mean of particulate phosphorus (% d.w.) for that year plus the mean concentration of the dissolved phosphorus. A mean of 50 mg suspended matter per litre leads to a content of 8500 tons in the open lake. Based on this figure it is possible to consider the quantity of sediment moved by stronger winds.

Table 4. Mean nutrient concentrations of the open lake, Wulka river and precipitation given in mg N resp. $P \cdot l^{-1}$.

	Open lake 1975	Open lake 1978	Wulka	Rain
NH_4^+	—	0.05	2.7	0.9
NO_3^-	0.18	0.10	2.4	0.7
P_{tot}	0.07	0.14	2.2	0.05
P_{diss}	0.03	0.04	1.3	—
PO_4^{3-}	0.01_5	0.02	1.1	0.01

Only a small amount of the nutrients leave the lake by the surface outflow, compared to the much larger quantities of nutrients monitored going into the lake. Due to the relatively high water input from precipitation the addition of inorganic nitrogen by rain is more important than the addition by the surface inflows, however the high phosphorus load of the inflows leads to a large influence of the inflows on the phosphorus addition. Therefore the type of junctions the various inflows have with the open lake is important in determining the phosphorus load reaching the lake. The Wulka water disperses into the reed belt while most of the other inflows reach the open lake by canals, although the small influence of the latter is caused by the unimportant water load. The contact of waste water with the reed belt retards the nutrient input into the open lake. The phosphorus increase over the last few years, in the area of the inflow of the River Wulka into the open lake, indicates an exhaustion of this part of the reed belt to take up further nutrients. The function of the reed belt acting as a reservoir for the land- and lake-wards inflowing nutrients hinders a faster eutrophication of the open lake.

Acknowledgement

The authors wish to express their thanks to the Air Hygienic Department of the Ministry of Health and Environment, International ECE Program for supplying rain samples and nitrogen data of precipitation.

References

Dokulil, M. 1975. Horizontal- und Vertikalgradienten in einem Flachsee (Neusiedler See, Österreich). Vhdl. Ges. Ökol. 177–187.

Golterman, H. L. 1975. Physiological Limnology. Development in Water Science 2, Elsevier Science Publishing Company, Amsterdam. 489 99.

Gunatilaka, A. 1978. Role of seston in the phosphate removal in Neusiedler See. Verh. Internat. Verein. Limnol. 20: 986–991.

Neuhuber, F. 1977. Die Nährstoffbelastung der Wulka und ihr Einfluß auf den Neusiedlersee. Biol. Forschungsinst. Burgenland Ber. 15: 38–44.

Neuhuber, F. 1978. Die Phosphorsituation des Neusiedlersees. Österreichische Wasserwirtschaft 30: 94–99.

Weiler, R. R. 1979. Rate of loss of ammonia from water to the atmosphere. J. Fish. Res. Board Can. 36: 685–689.

MECHANISMS CONTROLLING PHYTOPLANKTON PRIMARY PRODUCTION IN A TURBID SHALLOW LAKE

M. DOKULIL

Abstract

One of the most obvious features of Neusiedlersee is the high inorganic turbidity resulting from sediment stirred up from the bottom by wind action. Ample evidence has been presented previously on the influence of inorganic particle concentration on underwater light and on horizontal distribution of turbidity in relation to wind direction and currents. Since turbidity changes between less than 10 mg dry weight per liter and more than 800 mg, underwater PhAR-radiation is severely affected. It is to be expected therefore, that phytoplankton photosynthesis is strongly controlled by the wind and radiation regime of the lake. Other factors influencing primary production, like admixture of bottom algae, circulation of algae through steep light gradients and sedimentation of larger algal cells under calm weather conditions are of little significance.

Optimum photosynthesis of the phytoplankton is largely a function of biomass, as frequently observed elsewhere, but the increment of A_{opt} per unit change of biomass is surprisingly high which might be attributed to the large fraction of nannoplankton. Results from various other lakes support this conclusion. The variation of P_{opt} (the optimum photosynthesis per unit biomass) is explained up to 60% by temperature. The Q_{10} of P_{opt} is 2.3, comparable to results obtained elsewhere. P_{opt} varies independently from I_k (the irradiance denoting the onset of light saturation) which in turn seems to be temperature independent. The function A_{opt}/e explains 64% of the variation of the surface rate (A). Moreover the photosynthetic rate at 20 cm depth alone accounts for more than 86% of the variation in $\sum A$. The residual variance cannot be attributed to the light function I'_0/I_K nor to turbidity concentration.

The seasonal pattern of phytoplankton is published in: H. Löffler (Ed.), Neusiedlersee; The limnology of a shallow lake in Central Europe, pp. 203–321, Junk Publ., The Hague, Boston, London 1979. The complete text of this paper will be published together with the basic information in Hydrobiologia.

EFFECTS OF FOOD, PREDATION AND COMPETITION IN THE PLANKTON COMMUNITY OF A SHALLOW LAKE (NEUSIEDLERSEE, AUSTRIA)

A. HERZIG

Abstract

This survey shows the recent situation in the zooplankton of Neusiedlersee. Quantitative changes which have occurred in the last ten years are discussed on the basis of the relationship "phytoplankton – herbivorous zooplankton – potential predators". Since 1977 *Leptodora kindti* has been present in the plankton community; its phenology is described and also the possible impact on the filter feeders is discussed.

Introduction

Since 1968 a continuous survey of chemical and biological information about Neusiedlersee has been undertaken. For the zooplankton community of the open lake the results for the first six years can be found as a detailed qualitative and quantitative description in a monograph on Neusiedlersee (Löffler, 1979); in it relevant physical and chemical background data about the lake are described, as well as the development of the phytoplankton.

From the information existing about the zooplankton community it becomes obvious that two species – *Arctodiaptomus spinosus* and *Diaphanosoma brachyurum* – dominate the crustacean zooplankton, the cladoceran species being normally found during the summer months. The rotifers show a clear seasonal distribution, two to four species dominating in each season. Numbers and biomass of the crustaceans increase until 1971, afterwards they show a small decrease, but

never drop below the level of 1970; the reason for this is that there is an increase in nutrients and a correspondingly greater phytoplankton abundance. Numbers of the rotifers have fallen steadily since 1969 and show an inverse relationship with the crustacean numbers (especially in summer and winter); it is proposed that competition could be the reason for this.

Another factor is the decrease in abundance of the cyclopoid copepods (*Acanthocyclops robustus*, *Mesocyclops leuckarti*, *Thermocyclops crassus*) which happened in parallel to a decrease in macrophyte density. Older stages of the cyclopoids always show a clear preference for areas where macrophytes are numerous (Herzig, 1979).

The aim of this paper is to describe and to discuss the qualitative and quantitative development of the zooplankton of the open lake for the period 1975–1978 and to give a short description about the phenology of *Leptodora kindti*, a species which recently, 1977, appeared in the plankton of Neusiedlersee.

Materials and methods

During this study samples were collected with vertical net hauls (30 μm mesh size) through 1.55 m maximal depth. It resulted in integrated samples (50–120 l lake water filtered) without clogging of the net since most of the inorganic particles are smaller than 30 μm. The sample interval varied between 6–15 days in summer, 2 weeks in spring and autumn, and 3–5 weeks during winter. Details of the sampling, subsampling and counting techniques can be found in Herzig (1974, 1979).

In order to obtain a reasonable estimate of the lake's population, 8–20 horizontally distributed stations (Neusiedl–Illmitz/Mörbisch) have been sampled and counted so that a mean for the whole lake could be calculated; 95% confidence limits have been fitted to these calculated mean population estimates. All calculations were made using log-transformed values.

Biomass was determined by multiplying the mean numbers of animals (stages) per cubic metre by their corresponding mean dry weight. The dry weight per animal was derived from length/weight regressions which were established for the species (Bottrell et al., 1976).

Results and discussions

Quality and quantity of zooplankton – a continuation

Since 1975 Arctodiaptomus spinosus and Diaphanosoma brachyurum have dominated the crustacean plankton; in addition a clear decrease in the numbers of cyclopoids, Daphnia, Ceriodaphnia and Bosmina has become apparent. Until 1973 these Cladocera together represented annual means of 0.9–5.2% of the total crustacean numbers, but nowadays they rarely reach more than 1%. The same is true for the cyclopoids; occasionally percentages of 5% can be seen.

This means that until 1977 the crustacean plankton community consisted mainly of two herbivorous, filter-feeding species. In 1977 Leptodora kindti, an invertebrate predator, at least in the adult stage, appears in the plankton and since then has been increasing in numbers.

Another remarkable change can be seen in the rotifer populations, namely the "explosive" development of Rhinoglena fertöensis. It is a stenothermal species which appears from September/October to May. In the last seven years the relative abundance of this rotifer increased from 3% to nearly 100% of total rotifer numbers during the winter months (Herzig, 1980). In Fig. 2a the actual densities of the rotifers are given as means per season. It can be seen that the numbers have decreased since 1968 but from

1975 onwards there has been an upward trend. The dramatic increase in importance of Rhinoglena is reflected by the high winter mean values of more than 900 ind $\cdot$ l^{-1} (1976) and 500 ind $\cdot$ l^{-1} (1977) which are entirely caused by this species. It is also responsible for the increases in spring and autumn. This increase in population density can be explained by the good quality of the food supply and the parallel increase in quantity (Herzig 1980).

Total numbers and biomass of Arctodiaptomus spinosus and Diaphanosoma brachyurum is shown in Figs. 1 and 2. The number and biomass curves in Fig. 1 resemble those for the years 1975–1978, and show maximum total densities of 250–500 $\cdot$ 10^3 ind $\cdot$ m^{-3} and maximum total biomass estimates of 690–1100 mgdw $\cdot$ m^{-3}. Arctodiaptomus maxima are between 170 and 350 $\times$ 10^3 m^{-3} and 430–960 mgdw $\cdot$ m^{-3}, those for Diaphanosoma are between 96–200 $\cdot$ 10^3 ind $\cdot$ m^{-3} and 250–440 mgdw $\cdot$ m^{-3}. In Fig. 2 the mean values for the different seasons are compared for the period 1968–1978. As it can be seen, the highest values were recorded for 1971 and 1973 (Herzig, 1979), which were then followed by somewhat lower values. Nevertheless, the low numbers and biomass which occurred in 1968–1970 are not seen again.

In order to get some idea about possible reasons for the changes observed, phytoplankton biomass (expressed as Chla., annual means) was plotted against crustacean biomass (annual means) (Fig. 3). There is evidence that a greater food supply results in higher grazer biomass, the latter increasing more than the algal biomass which is controlled by the zooplankton until 1971. This tendency lasted until 1971/72 and since then has been followed by a decrease in algal and crustacean biomass. From 1973 until 1975/77 algal biomass increased although the mean crustacean biomass continued to decline, whereas in 1978 zooplankton, as well as phytoplankton showed an increase.

The crucial point in this relationship is the fact that the decreasing crustacean biomass occurs at a time when the food quantity is increasing. Most probably this has to be seen in relation to the development of planktivorous fish, of which Pelecus cultratus seems to be the most important. From gut analysis (Herzig unpublished results) it

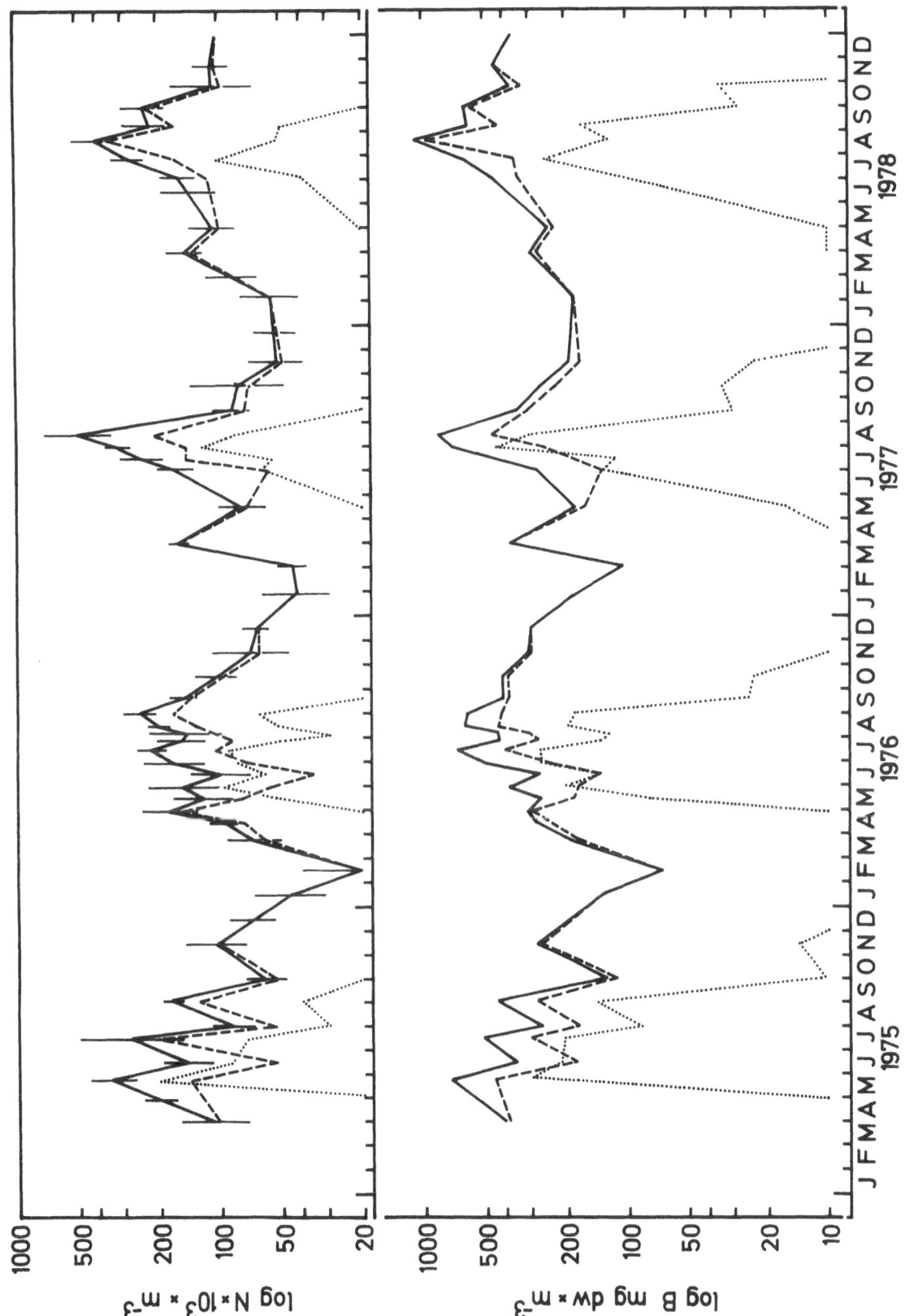

Fig. 1. Crustacea – abundance (N) and biomass (B). Vertical lines indicate the 95% confidence intervals. ———— total Crustacea, ———— Arctodiaptomus spinosus, ·········· Diaphanosoma brachyurum.

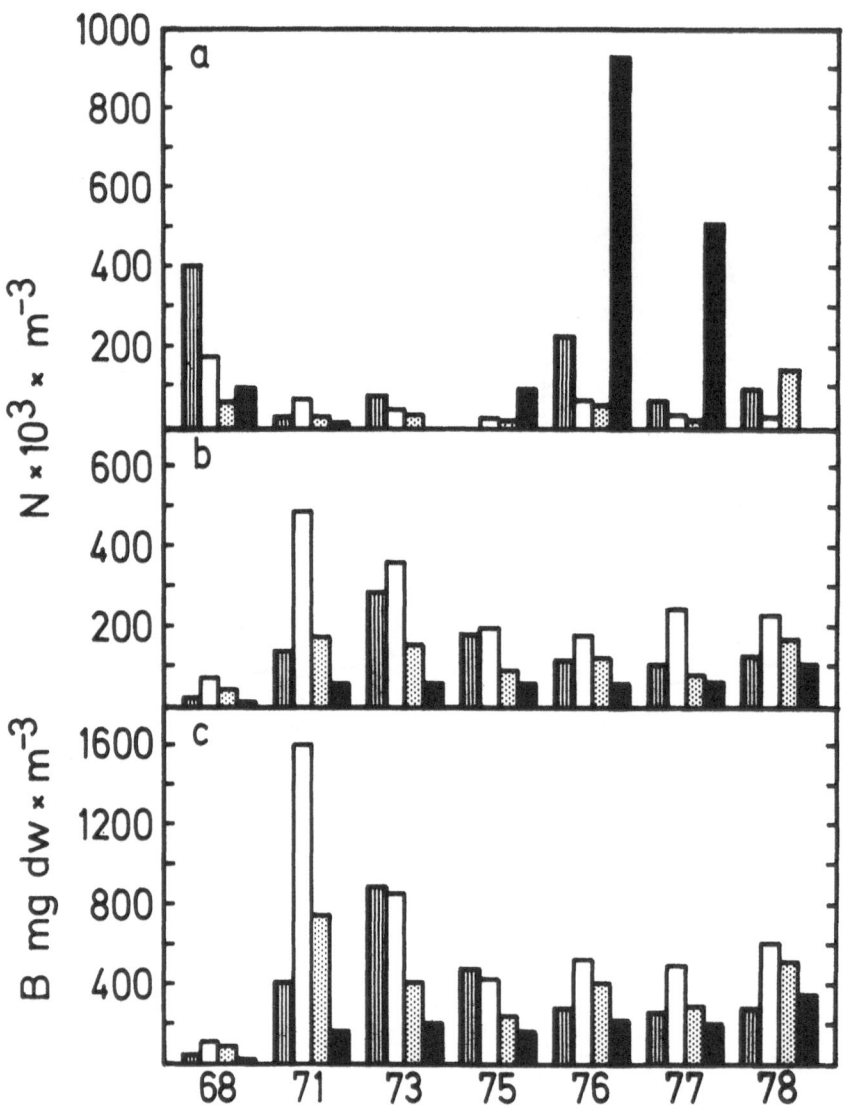

Fig. 2. Seasonal means of (a) rotifer abundance, (b) crustacean abundance, (c) crustacean biomass.
⬛ spring, ☐ summer, ⊞ autumn, ■ winter.

48

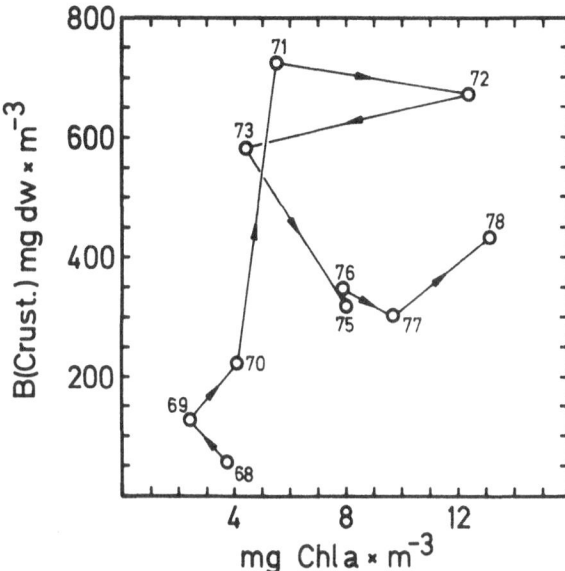

Fig. 3. Relation between phytoplankton biomass (mg Chl$a \cdot$ m^{-3}) and crustacean biomass (mg dry weight$\cdot$ m^{-3}) (Phytoplankton data: DOKULIL 1979 and unpublished results).

becomes obvious that *Pelecus* was nearly exclusively feeding on crustacean plankton, only occasionally "Anflug" could be found in the guts. The abundance of this fish is also increasing. According to Hacker & Meisriemler (unpublished results) *Pelecus cultratus* was not caught by gill nets (#20 mm; length: 25 m, height: 1.5 m) until 1972. Since 1973 they frequently occur in catches (35–70 individuals, max.: >100 individuals; exposure time: 2 hours) and represent up to 90% of the total catch.

The current idea is that the increase in zooplankton (from 50 mgdw $\cdot$ m^{-3} to 725 mgdw $\cdot$ m^{-3}) has favoured the development of *Pelecus cultratus*. With the increase in fish abundance the zooplankton populations become depressed and between June and September are controlled by *Pelecus*; also the unknown quantities of the fry of all species of fish should not be ignored as potential consumers of zooplankton.

For the future it will be of interest to gather quantitative data for *Pelecus cultratus* (some information is already existing for 1978) and fish fry, as well as experimental information about the feeding habits of these animals in order to quantify the impact of planktivorous fish on the zooplankton.

The development of Leptodora kindti and the puzzling future of zooplankton

Leptodora was detected for the first time in mid June 1977 and since then this species has become more and more abundant. The growing season lasts from April/May until end of November. Maximum numbers and biomass were reached in July (1978) or September (1977), at the same time as the highest subitaneous egg numbers were recorded (60–100 eggs $\cdot$ m^{-3}). The first males appeared by mid-September and this coincided with the production of resting eggs. The percentage of males varied between 40 and 60% of total numbers. Hatching of the resting eggs started at the end of April or the beginning of May when water temperatures exceeded 10°C.

Most of these features of the phenology are very similar to those obtained for populations of other shallow temperate lakes (e.g. Sanctuary Lake/Pymatuning Reservoir, Cummins *et al.*, 1969; Lake Balaton, Sebestyén 1931, 1960).

The maximum densities are close to the values given by Sebestyén (1960 – Lake Balaton), Odermatt (1970-Lauerzersee), de Bernardi & Soldavini (1976 – Lago Mergozzo), de Bernardi & Canali (1975 – Lago Maggiore), which are in the range 60–100 individuals $\cdot$ m^{-3}. Extremely high maximum densities are recorded by Cummins *et al.* (1969 – Sanctuary Lake/Pymatuning Reservoir) or Hillbricht-Ilkowska & Karabin (1970 – Mikolajskie) reaching 1200–4200 individuals$\cdot$m^{-3}.

Now the question arises, what are the possible effects of *Leptodora* on a zooplankton community? In many situations the *Daphnia* populations are known to be controlled by *Leptodora*, e.g. Base Line Lake-Hall (1964), Canyon Ferry Reservoir-Wright (1965), Mergozzo-de Bernardi & Canali (1975); between 25–35% of the *Daphnia* population can be removed by *Leptodora*. In Lake Mikolajskie *Leptodora* may consume 15–43% of the cladoceran production (July/August) and the adult stages are most intensively eaten (Hillbricht-Ilkowska & Karabin, 1970). Mordukhai-Boltovskaya (1958) found in Rybinsk Reservoir *Leptodora* predating on *Bosmina* and *Ceriodaphnia*, whereas in Lago Mergozzo *Diaphanosoma brachyurum* is the preferred prey item (de Bernardi & Soldavini 1976). Summing up, it seems that *Leptodora* is an opportunistic

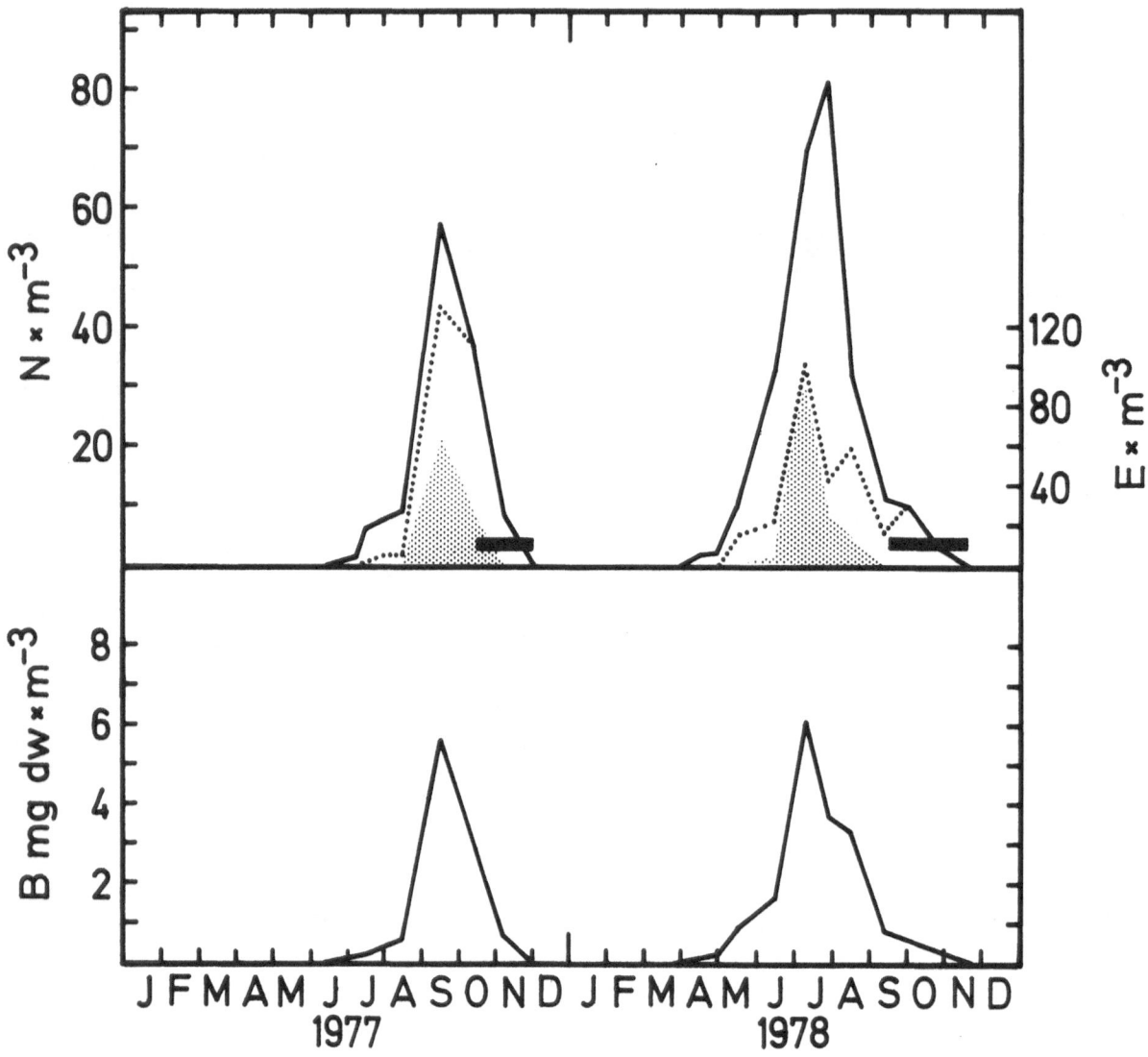

Fig. 4. Leptodora kindti - abundance (N) and biomass (B); shaded areas – number of subitaneous eggs; solid line – total numbers/biomass, dotted line – number/biomass of adults; horizontal bars indicate occurrence of males.

predator feeding on the most readily available prey item. Assuming this feeding strategy, *Leptodora* will feed in Neusiedlersee on *Diaphanosoma brachyurum* in the summer months. Applying the results of Cummins *et al.* (1969) for the ingestion of *Leptodora* to the time of its maximum occurrence in Neusiedlersee, 0.6–1.6% of the *Diaphanosoma* biomass could be eliminated per day. Applying the results of the experiments of Hillbricht-Ilkowska & Karabin (1970) the elimination of prey biomass would be 1.1–1.6% per day. Even from these approximate calculations,

based on several assumptions, it can be seen that *Leptodora* will play an important role in the planktonic food chain of Neusiedlersee. In addition one should bear in mind that although the actual densities of *Leptodora kindti* are not very high at present, they show an increasing trend.

On the other hand *Leptodora* also serves as food for fish fry and planktivorous adult fish. Since 1977 *Leptodora* has been found in the guts of *Pelecus cultratus* (Herzig unpublished results) and it is of interest whether this fish can control the growth of this invertebrate predator. Hence the fate of *Leptodora* in Neusiedlersee will be of interest in future investigations.

Summary

(1) The changes in abundance and biomass of zooplankton are described and discussed for the years 1975–1978. In addition a short comparison is given for the years 1968–1978, where an attempt is made to show the quantitative development of phyto- and zooplankton in the last decade.

(2) A short description is given of the phenology of *Leptodora kindti*, which first appeared in Neusiedlersee during 1977. Data on this species is also compared with results from the literature.

(3) The impact of *Leptodora*, as an invertebrate predator, on *Diaphanosoma brachyurum* is discussed on the basis of both published and Neusiedlersee data.

Acknowledgements

The author is indebted to Dr. U. Einsle for his help with cyclopoid systematics and Dr. M. Dokulil for providing unpublished phytoplankton data. This survey was done in the scope of a MaB-study on Neusiedlersee and hence was sponsored by it.

References

Bottrell, H. H., Duncan, A., Gliwicz, Z. M., Grygierek, E., Herzig, A., Hillbricht-Ilkowska, A., Kurasawa, H., Larsson, P. & Weglenska, T. 1976. A review of some problems in zooplankton production studies. Norw. J. Zool. 24: 419–456.

Cummins, K. W., Costa, R. R., Rowe, R. E., Moshiri, G. A., Scanlon, R. M. & Zajdel, R. K. 1969. Ecological energetics of a natural population of the predaceous zooplankter Leptodora kindtii Focke (Cladocera). Oikos 20: 189–223.

DeBernardi, R. & Canali, S. 1975. Population dynamics of pelagic cladocerans in Lago Maggiore. Mem. Ist. Ital Idrobiol. 32: 365–392.

DeBernardi, R. & Soldavini, E. 1976. Long-term fluctuations of zooplankton in Lago Mergozzo, Northern Italy. Mem. Ist. Ital. Idrobiol. 33: 345–375.

Dokulil, M. 1979. Seasonal pattern of phytoplankton. In: H. Löffler (ed.) Neusiedlersee, the limnology of a shallow lake in Central Europe. Dr. Junk, The Hague, 203–231.

Hall, D. J. 1964. An experimental approach to the dynamics of a natural population of Daphnia galeata mendotae. Ecology 45, 94–112.

Herzig, A. 1974. Some population characteristics of planktonic crustaceans in Neusiedlersee. Oecologia 15, 127–141.

Herzig, A. 1979. The zooplankton of the open lake. In: H. Löffler (ed.) Neusiedlersee, the limnology of a shallow lake in Central Europe. Dr. Junk, The Hague, 281–335.

Herzig, A. 1980. Ten years quantitative data on a population of Rhinoglena fertöensis (Brachionidae, Monogononta). Hydrobiologia, in press.

Hillbricht-Ilkowska, A. & Karabin, A. 1970. An attempt to estimate consumption, respiration and production of Leptodora kindtii (Focke) in field and laboratory experiments. Pol. Arch. Hydrobiol. 17 (30) 1/2: 81–86.

Löffler, H. (ed.). 1979. Neusiedlersee, the limnology of a shallow lake in Central Europe. Dr. Junk, The Hague, 543 pp.

Mordukhai-Boltovskaya, E. D. 1958. Preliminary notes on the feeding of the carnivorous cladocerans Leptodora and Bythotrephes. Dokl. Akad. Nauk. SSSR, Biol. Sci. Sect., 122: 828–830.

Odermatt, J. M. 1970. Limnologische Charakterisierung des Lauerzersees mit besonderer Berücksichtigung des Planktons. Inaugural-Dissertation, Universität Zürich. 575 pp.

Sebestyén, O. 1931. Contribution to the biology and morphology of Leptodora kindtii Focke (Crustacea, Cladocera). Arb. Ung. Biol. Forschungsinst. 4: 151–170.

Sebestyén, O. 1960. Quantitative plankton studies of Lake Balaton. X. Notes on the distribution of Leptodora kindtii (Focke). Ann. Inst. Biol. (Tihany) Hung. Acad. Sci. 27: 131–138.

Sebestyén, O. 1960. On the food niche of Leptodora kindtii Focke (Crustacea, Cladocera) in the open water communities of Lake Balaton. Int. Rev. ges. Hydrobiol. 45, 2: 277–282.

Wright, J. C. 1965. The population dynamics and production of Daphnia in Canyon Ferry Reservoir, Montana. Limnol. Oceanogr. 10: 583–590.

WASTEWATER TREATMENT AND PHOSPHORUS LOADING OF NEUSIEDLERSEE

H. FLECKSEDER, E. RUIDER & G. SPATZIERER

Abstract

The loading of tot P to Neusiedlersee in 1977 was 40% from point-sources and 60% from non-point-sources. In order to reduce the loading from point-sources, simultaneous precipitation with Fe^{+2} seems at present to be the most attractive method and has been in use since the beginning of 1979. The estimates of the loading from non-point-sources included here should aid future decisions on which management policy to pursue in the future.

Introduction

Neusiedlersee is endangered by

- local algal blooms (O_2-balance)
- lakewide algal blooms (O_2-balance)
- increased aging.

The cause of these problems in Neusiedlersee has not yet been fully identified, but according to Dokulil, Herzig (1976) and Neuhuber (1976, 1978), the increased input of phosphorus can be looked upon as a primary reason. Beyond this, a change in utilization and handling of the reeds also seems, in our view, to play a contributary role. The increased input of phosphorus has primarily been associated with insufficient wastewater treatment, with direct input from tourism and agricultural discharges in the broadest sense. The quick installment of a basin transfer system for point source discharges has been looked upon as a "complete solution to Neusiedlersee's problems"

and has therefore been promoted by some environmentalists as well as the construction industry, thus questioning the existing concept of wastewater treatment. This concept is the building of sewers and wastewater treatment plants with a high degree of efficiency for loads high in organic carbon. It was felt that the reeded area served as a natural phosphorus purge and therefore simultaneous precipitation of phosphorus (i.e. the addition of ferrous salts into the activated sludge process) was not introduced up until 1977 at certain locations. In order to judge which policy of wastewater treatment should be adopted in the future, the authorities in charge of water quality within the federal state of Burgenland asked for an investigation. The following paper covers part of this work.

Procedure of study

As in most similar cases, base data on nutrient inputs were missing. However, decisions had to be taken and the time available to measure them in the field was limited. In an applied study like this one the following position and procedures were felt to be accurate (W. Stumm, 1971):

- The increased discharge of matter into waters is in most cases caused by human activity.
- The state of "water quality", therefore, is in most cases caused by human activity in the drainage basin of a water system.
- Applied to Neusiedlersee, one can say that the

trophic state of the lake is determined by the past and present discharge of matter from households, trade, industry, tourism, sealed surfaces, animal breeding, the washout and the erosion from agricultural soils, the input via air, the net input from birds but also by a change in utilization of the reed.

- Therefore, in order to assess the input to and the output from the lake of, for example, phosphorus one has to quantify every "(human) activity" by a number designating how many-times the "activity" is occurring as well as the specific value of the input or output.
- As far as possible the indirect method (activity times specific load) should be substantiated by direct measurements.
- The method of analogy (i.e. indirect method) allows a more easily arrived at prediction of past and future states of loading, but that still remains only an estimate of the true state. The proof of the method of analogy is its application to presently discharged and measured loads.
- As can be anticipated administrative as well as technological countermeasures have to be implemented to cope both with non-point-source as well as point-source loads.

The question on the permissible load of phosphorus to Neusiedlersee and its influence on the processes occurring within the lake (spatial and temporal distribution of P and its effect on phyto- and zooplankton) as well as the question of whether the reduction in P will start a recovery of the lake was, as true limnological work, not part of the investigation asked for. Input of P to and (possible) output from the lake were considered as for a black box with limits of the land-water boundary, on the inland side of the reed belt, some meters above the top of the reeds, and some meters below the solid-water interface at the bottom of the lake.

Application to Neusiedlersee

Balance

$$Input - Out = Sink$$

The task is to thoroughly assess all existing inputs and outputs and to forget nothing.

Input

Method of anology

Estimates for "activities"

General remarks

The "method of analogy" has to start from estimates of "activities" and from estimates of the "specific loads" in order to obtain estimates for the loads discharged. These final estimates are naturally only as good as the basic ones. This method is, therefore, not free from errors, but it appears that certain errors cancel out and that the estimate for the overall load seems realistic.

Population

(as a measure for loads in sewage)

Dimension: People	Permanent residents	Summer peak
In the future within Austria	84 600	205 000
In the future within the whole drainage basin	91 700	215 000

Sealed surface

In the future within Austria 1700 ha. Within the whole drainage basin 1830 ha.

Industry

As the industry in the drainage basin is based on agricultural raw materials (fruit and vegetables canning, beet sugar processing) and as the loads from these are known by direct measurement, they are dealt with elsewhere.

Agriculture

(a) Animal breeding
Cattle 11 600 ($\downarrow$), pigs 63 000 ($\uparrow$), chicken 278 000 ($\downarrow$), turkey 17 300 ($\downarrow$), horses 817 ($\leftrightarrow$)
(b) Utilization of soils
Fields 50.7% ($\downarrow$), gardens 0.5% ($\leftrightarrow$), vineyards 10.4% ($\uparrow$), fruit growing 0.3%, meadows 8.5%, forest 24.6% remainder 5.0%. Roughly 62% of the soils are more or less heavily fertilized (fields, gardens and vineyards). The drainage area to the lake is in Austria 1020 km², the overall drainage area to the lake is 1100 km². Tendencies: raising ($\uparrow$), equal ($\leftrightarrow$), falling ($\downarrow$) for 1976 over 1973.

Transport via air

Liquid water surface of 285 km^2 taken into account.

Subsurface inflow

$8 \cdot 10^6$ m^3/a taken into account (Kopf, 1967).

Bathing

60 000 tourists for 90 days plus 115 000 bathers (tourists plus local people) for 8 weekends makes in all $7.24 \cdot 10^6$ person-bathing-days (Estimate based on Statistics).

Net input from birds

A large number of birds live, during spring and summer, in the reedbelt, which covers an area of roughly 140–150 km^2. The question arose as to whether there is a net input from the daily migration of birds or not, but it was impossible to answer this question. Hamm (1975) reports on the eutrophication of small lakes (gravel pits) in Bavaria as the result of the excretion of birds alone. This point is dismissed in the quantification, but not forgotten as a problem.

Estimates for "specific leads" population

4.0 g tot P/person · day was used as an estimate for the "gross load".
At four plants within the drainage basin, COD, BOD$_5$, TOC, tot N and tot P were observed for two weekends by continuous sampling. It was impossible to assess the population present, but when normalizing the COD to 100 and the BOD$_5$ to 60, the following ratios were obtained:

100 COD : 33.6 TOC : 44.5 BOD$_5$: 11.4 tot N : 3.01 tot P

and

143.5 COD : 45.5 TOC : 60 BOD$_5$: 15.3 tot N : 4.07 tot P.

In the inflow to a different plant in the western part of Austria, such observations lasted for six months and showed ratios of

100 COD : 29.9 TOC : 52.4 BOD$_5$: 9.1 tot N : 2.7 tot P

and

115 COD : 34.4 TOC : 60 BOD$_5$: 10.2 tot N : 3.1 tot P.

Both observations are in fair agreement with the estimate, and the value chosen "allows" for some further increase in this specific figure.

This "gross load", when linked to the sewer, has to be reduced. One way to a fair reduction is the substitution of phosphorus by different agents in detergents, but this will only help to obtain some 50% reduction and cannot be introduced on such a small scale. Technological possibilities with the existing treatment plants are simultaneous precipitation (i.e. the addition of ferrous sulfate into an activated sludge plant) at various doses of Fe/P per weight, the expansion of two big treatment works with contact filtration (i.e. precipitation of dissolved and colloidal P and separation by rapid sand filtration) and basin transfer. The removal of tot P was rated as 25% efficiency for normal designs of activated sludge plants, as 90% for low Fe^{+2}/P ratios (2–3) and as 97% for high Fe^{+2}/P ratios (4–5), see v.d. Emde and Spatzierer (1978). Simultaneous precipitation followed by contact filtration was rated as 99% (Boller and Kavanaugh, 1977), although Gujer and Boller (1978) report only 97%. Basin transfer was rated as 100%.

Simultaneous precipitation can be quickly implemented and requires predominantly operating costs, whereas contact filtration needs a longer timespan to be implemented. Construction work for basin transfer will certainly be longest. Most of the working experience exists with simultaneous precipitation, whereas contact filtration is a relatively new process. Should problems arise in the longer run with simultaneous precipitation (e.g. shortage of precipitant), biological P-removal (e.g. Simpkins and McLaren, 1978, and Arueste, 1979), should be tested as a possible solution.

Sealed surfaces

Averages from published data (Göttle, 1976; Roberts *et al.*, 1976; Sartar, 1974; Söderlund, 1972; Weibel, 1969) yield an estimate of 2.5 kg tot P/ha.a. By providing adequate storage capacity in the combined sewer system it was assumed that 90% of the tot P in the runoff from storms can be taken to a treatment plant with only 10% loss (compare Krauth, 1971).

Agriculture

In "Wege und Verbleib des Phosphors in der Bundesrepublik Deutschland" (Anonymous, 1978) it is estimated that there is a contribution of 2–3% of the excreta of animals. This study is based on a contribution of 2%, which again is classified as non-point-source pollution. Also in "Wege und Verbleib des Phosphors in der Bundesrepublik Deutschland" specific data of 37 g P/cattle · d, 9.4 g P/pig · d and 0.5 g P/chicken · d are reported. Horses were estimated to be equal to cattle, turkeys to hold 2.5-times the amount of chickens. A literature survey on the specific load of P eroded by water from agriculturally used soils (Hamm, 1976) shows figures which range from 0.03 to 2.0 kg tot P ha^{-1} a. A very thorough study for Central European conditions (Gächter and Furrer, 1972) allows an estimate of 0.2 kg P ha^{-1} a^{-1} for reasonably sloped regions with roughly 70% of agriculturally used land in the drainage basin.

Transport via air

A literature survey on the specific load of tot P transported via air (Hamm, 1976) revealed that this figure can vary strongly (0.02 to 1.12 Kg ha^{-1} a^{-1}). In a very thorough study Bernhardt and Wilhelms (1978) reported 0.97 kg tot P ha^{-1} a^{-1} for Wahnbachtalsperre, which is situated in a forested region, but not far away from industry. Peukert and Panning (1975) reported even 2.4 kg tot P ha^{-1} a^{-1} for a water supply reservoir in the GDR, and Sampl (1978) 1.6 kg tot P ha^{-1} a^{-1} for Ossiachersee in Carinthia/Austria.

Agricultural centers in Austria and Hungary are not only situated in the drainage basin of Neusiedlersee, but also in a bigger area towards the west, north and east. Measurements of N and P in the water collected with non-recording rain gauges were started in May 1978 for locations at Rust and Podersdorf. Very strong fluctuations in monthly concentrations (total P, total P after sedimentation, PO$_4$–P) as well as in monthly loads were observed. For Rust, the averages were 2.0 kg tot P ha^{-1} a^{-1} (May 78–July 79), 0.84 kg tot P after sedimentation/ha a (August 78–July 79) and 0.73 kg PO$_4$–P ha^{-1} a^{-1} (May 78–July 79), whereas for Podersdorf they were 2.4 kg tot P ha^{-1} a^{-1} (May 78–June 79), 0.90 kg tot P after sedimentation/ha. a (August 78–June 79) and 1.1 kg PO$_4$–P ha^{-1} a^{-1} (May 78–June 79). As it is quite unlikely that dissolved phosphate is in the air in appreciable amounts and as we believe that most of the phosphorus transported via air originates from wind erosion, we believe that the estimate presently to be applied should be 2.0 kg tot P ha^{-1} a^{-1}.

Subsurface inflow

In the literature (e.g. "Wege und Verbleib des Phosphors in der Brd"), concentrations from 5 · · · · 200 μg l^{-1} tot P are reported for non-acidic soils. An "average" value of 30 μg l^{-1} was selected for this study.

Bathing

Schulz (1976) measured a specific tot P-figure of 95 mg/person bathing day and this value is applied for this work.

Estimated loads
Point-sources

The point-source load comprises the discharges from households, trade, industry, stored surface runoff and from the visits of toursits. These figures hold for the Austrian drainage basin only and have to be looked upon as a load potential after the described treatment has been applied and as estimates for future conditions. The estimate for the Hungarian side is a gross potential of 14.0 t tot P a^{-1} which certainly is being and will be further reduced by wastewater treatment. It has to be stressed, however, that these loads after treatment hold only as long as the point-source discharge remains in the sewer and doesn't enter the recipient directly. The authorities are fully aware of these problems and have started a thorough control program.

	t tot P a^{-1}
a) No treatment at all	199
b) Biological treatment only	149
c) Simultaneous precipitation at low Fe/P-doses	

d) Simultaneous precipitation at
low Fe/P-doses + contact
filtration at regionalized
plants

e) Simultaneous precipitation
at high Fe/P-doses at all
plants   6

f) Biological treatment and
basin transfer   0

This table, however, should not be interpreted in such a way that in the past amounts such as 149 or even 199 t tot $P\,a^{-1}$ have in fact entered the lake. Both these values are included as an estimate of the "gross potential" before treatment for P-removal is applied and in a case where *all* the population is assumed to be sewered. However, this has not been so in the past and the discharge to sewers or recipients has been quite well correlated with the building of wastewater treatment plants.

Non-point-sources

These figure roughly hold for present conditions

	t tot $P\,a^{-1}$
Via air	57.0
Eroded from land by water	22.0
From animal breeding	10.0
From not collected storm runoff (90% reduction assumed)	0.4
From bathing	0.7
Subsurface inflow	0.2
Estimated total	90.0

for the whole lake (drainage basin plus lake area).

Direct measurements

Limited measurements at the inflows to the lake on the Austrian side allowed "measured estimates" for tot P of roughly 63 t for 1976 and roughly 52 t in 1977 at dry weather periods. It is felt that these figures are primarily associated with point-source loads.

Comparison of direct measurements and the method of anology

The present population living on the Austrian side is 91 600 people and they contribute the biggest share to point-source pollution. If we assume 3.5 g tot P/PE.d to be correct and if we further estimate that 65% of these people have been sewered in 1976 and 1977, the discharge to sewerage is then 76 t tot $P\,a^{-1}$. With a P-reduction of 25% by activated sludge and with 80% of the sewered population serviced by biological treatment this gives a yearly P-load of 61 t tot P. This is in good agreement with the measurements.

Output

The output from the lake includes losses down the outflow, from the fish harvested, and from the reed cut.

Estimates are	t tot $P\,a^{-1}$
– for the outflow $10 \cdot 10^6\,m^3\,a^{-1}$ at $200\,\mu g$ tot $P\,l^{-1}$ (An outflow of $10 \cdot 10^6\,m^3\,a^{-1}$ is the value determined by Kopf 1967, but according to unpublished estimates $25 \cdot 10^6$–$30 \cdot 10^6\,m^3\,a^{-1}$ flow out. The value of $200\,\mu g\,l^{-1}$ is somewhat higher than the average of $150\,\mu g\,l^{-1}$ reported for the free water surface in 1976 (Neuhuber, 1978). The outflow is situated in a reeded area where silt settles out, but where an internal loading seems to occur, too).	2.0
– for the fish harvested $420\,t$ fish a^{-1} with 9% dry substance for the fishbone and of that 9.5% as P	3.6
– for the reed cut (150 km^2 of reed considered)	
– under the present Austrian usage of $0.15\,t$ reed $ha^{-1}\,a^{-1}$ at 0.13‰ P in dry substance (winter cut)	0.3
– at a useful economic yield of $5.0\,t$ reed $ha^{-1}\,a^{-1}$ with winter cut	9.8
– at a maximal cut of $15\,t$ reed ha^{-1} 1.0‰ P of dry substance directly after the growing season (According to R. Maier, cited in Löffler, 1974, the standing crop of the reed is $120\,t\,ha^{-1}$. Every year, $30\,t\,ha^{-1}$ grow anew and die away. If reed is harvested for construction purposes, it has to be cut in winter and the	225

maximal economically interesting yield is 5 t ha^{-1}. The P-content in the reed shifts between the rhizome and the stem including leaves. Above the soil it is high during the growing season and directly after it. First determinations in July 1979 showed, as an average of 4 samples, 1.0‰ of P on a dry weight base. In winter, however, the P-content in the leaves and the stem is strongly reduced. At the end of February 1979, an average of 0.13‰ on a dry weight base was obtained for 3 samples).

Discussion

On the *input side* it is evident that
- no wastewater treatment or biological treatment only do not present viable alternatives
- with such a shallow lake and with the small ratio of drainage area versus lake surface, the transport via air contributes strongly to the non-point source load
- simultaneous P-precipitation at low doses is at least required
- any choice on the treatment process(es) to be selected must consider the additional cost versus damage remaining and also the time required to implement the solution
- not only wastewater treatment, but also administrative measures in agriculture should be considered.

On the *output side* only the harvesting of reeds quite soon after their growing season and disposing of them away from the lake seems to be an attractive way to reduce the phosphorus pool in the lake. How that can be done, what direct effect and what side effects this will have remains to be answered.

The *yearly differential balance of phosphorus* will remain positive as long as the output cannot be increased over the input. The future state of the open-water seems to depend to quite a large extent on the processes occurring within the reedbelt and the subsequent exchange with the open waters. From this it seems that *applied limnological research* should concentrate on

- *assessing the cycling of P between the reedbelt and the open waters*
- *observing the non-point-source input via air and from erosion by water*
- *obtaining administrative measures that are legally formulated and applied to non-point sources.*

Point-source loads are since Spring 1979 taken care of by *the introduction and thorough control of simultaneous precipitation* and by speeding up construction work for the biological treatment plants still to be built.

Basin transfer has been ruled out due to the long period for construction and the high cost involved.

5. Acknowledgements

The authors would like to express their thanks to their former colleague and friend W. Stalzer, Amt der Burgenländischen Landesregierung, for his excellent cooperation when trying to solve the questions raised.

References

Anonymous (1978). "Phosphor–Wege und Verbleib des Phosphors in der Bundesrepublik Deutschland". Verlag Chemie/Weinheim, xviii + 285 p.

Arueste, G. (1979). "Advanced Municipal Wastewater Treatment by a Single Stage Activated Sludge Plant". Paper presented at the IAWPR-Vienna-Workshop, September 3–7, 1979, to be published in Prog. Wat. Techn.

Bernhardt, H. and Wilhelms, A. (1978). "Ergebnisse der Untersuchungen der Niederschläge im Bereich der Wahnbachtalsperre auf ihren Phosphorgehalt". KA 25: pp. 75 f.

Boller, M. A. and Kavanaugh, M. C. (1977). "Contact Filtration for Additional Removal of Phosphorus in Wastewater Treatment". Prog. Wat. Techn. 8: pp. 203 f.

v. d. Emde, W. and Spatzierer, G. (1978). "Das Klärwerk Zellerbecken". ÖWW 30: pp. 85 f.

Dokulil, M.: Personal communication.

Gächter, R. and Furrer, O. J. (1972). "Der Beitrag der Landwirtschaft zur Eutrophierung der Gewässer in der Schweiz. I. Ergebnisse von direkten Messungen im Einzugsgebiet verschiedener Vorfluter. II. Einfluß von Düngung und Nutzung des Bodens auf die Stickstoff- und Phosphormengen im Wasser". Schweiz. Z. Hydrol. 34: p. 41–93.

Göttle, A. (1976). "Die Beschaffenheit und die Behandlung von Regenwasser im Trennverfahren". Report Nr. 12 of the Institute for Civil Engineering V, (in German), TU München, pp. 1 f.

Gujer, W. and Boller, M. (1978). "Basis for the Design of Alternative Chemical-Biological Wastewater Treatment Processes". Prog. Wat. Techn. 10: pp. 741–758.

Hamm, A. (1975). "Chemisch-biologische Gewässeruntersuchungen an Klein- und Baggerseen im Großraum von München im Hinblick auf die Bade- und Erholungsfunktion". Münchn. Beitr. Z. Abw.-, Fisch- u. Flußbiol. 26: pp. 75–109. Oldenbourg-Verlag.

Hamm, A. (1976). "Zur Nährstoffbelastung von Gewässern aus diffusen Quellen: Flächenbezogene P-Abgaben – Eine Ergebnisund Literaturzusammenstellung", ZfWAF, 9: pp. 4 f.

Herzig, A. (1976). "Das Zooplankton des Neusiedlersees (Ein Überblick über die Jahre 1950–1975)". Vortrag and der 2. Neusiedlerseetagung, 23./24.9. 1976 in Illmitz.

Krauth, K.-H. (1971). "Der Abfluß und die Verschmutzung des Abflusses in Mischwasserkanalisation bei Regen". R. Oldenbourg-Verlag, 251 p.

Kopf, F. (1967). "Die Rettung des Neusiedlersees". ÖWW 19: pp. 139 f.

Löffler, H. (1974). "Der Neusiedlersee". Molden-Verlag, Vienna 175 pp.

Neuhuber, F. (1976). "Phosphoruntersuchungen im Neusiedlersee". Vortrag an der 2. Neusiedlerseetagung, 23./24.9.1976 in Illmitz.

Neuhuber, F. (1978). "Die Phosphorsituation des Neusiedlersees". ÖWW 30: pp. 94 f.

Peukert, V. and Panning, C. (1975). "Einfluß anorganischer Luftverunreinigungen auf die Wasserbeschaffenheit von Trinkwassertalsperren". Acta hydrochim. hydrobiol. 3: pp. 545 f.

Roberts, P. V. et al. (1976). "Schmutzstoffe im Regenwasser einer städtischen Trennkanalisation". Gas-Wasser-Abwasser 56: pp. 672 ff.

Sampl, H. (1978). "Das Kärntner Institut für Seenforschung", ÖWW 30: pp. 78 f.

Sartar, J. D. and Boyd, G. B. (1972). "Water Pollution Aspects of Street Surface Contaminants". Contract No. 14-12-921, Project 11034 FUJ, for US-EPA, November 1972.

Schulz, L. (1976). "Nährstoffeintrag in Seen durch Badebetrieb". Vortrag an der Tagung deutschsprachiger Limnologen der IVL, Oktober 1976 at Innsbruck/Austria.

Simpkins, M. J. and McLaren, A. R. (1978). "Consistent biological phosphate and nitrate removal in an activated sludge plant". Prog. Wat. Techn. 10: Nr. 5. p. 433–442.

Söderlund, G. and Lethinen, H. (1972). "Comparison of Discharges from Urban Stormwater, Runoff, Mixed Storm Overflow and Treated Sewage". Proceedings of the 6th International Conference on Water Pollution Research held at Jerusalem in 1972, pp. 309 ff. Pergamon Press.

Stumm, W. (1971). "Einfache Modelle im Umweltschutz: Der Mensch und die hydrogeochemischen Kreisläufe". Jahrbuch Vom Wasser, Vol. 38: p. 1–16.

Weibel, S. R. (1969). "Urban Drainage as a Factor in Eutrophication" in "Eutrophication-Causes, Consequences, Correctives", pp. 383 f. National Academy of Sciences, USA.

3. LAKE BALATON, HUNGARY

BACKGROUND DATA

Latitude: 47°3'50"–46°42'6"N
Longitude: 17°14'58"–18°10'28"E
Altitude: 104.5 asl.
l (km)   77
b (km)   15 (max)–9 (mean)
L (km)   200
L_D 2.26
Name of the main tributary   Zala.
Average inflow m³/sec.   12 (Zala = 6)
Average outflow m³/sec.   Sió 10
Theoretical retention time   3–8 year

origin: tektonic fault +
    erosion
z (m)   11
$\bar{z}$ (m)   3.3
$\bar{z}/z$ 0.3
V (km³)   1.8
A (km²)   600
A' (km²)
$A' : A$

Geological characteristics:

Its age is about 18,000 years. In its surroundings there are layers of the Pannonian Sea, with several basaltic inselbergs. Along the northern shore are thick silt deposits, while along the south shore a sand bank is formed.

Climatic conditions:
Average monthly temperature

J	F	M	A	M	J	J	A	S	O	N	D	y.
1	1	6	11	18	22	24	23	20	14	8	3	11.3

Average precipitation/year 640 mm
Average sunshine duration 2000 hrs/y
Main wind direction(s) NW (E)
Evaporation per year 840 mm

Ice cover (days) 60 days
Average radiation/year 115 kgcal/m²
% of calm days 25
$\bar{V}_{wind}$/year 3.2 m/sec.

Cultural geography and demography:
Land usage of catchment area (%)
Industrial   3
Agricultural   75
Meadows   3

Usage of lake water including recreation activities:
Commercial fisheries, sport fishing, transport, drinking water purposes, cooling water for industry; receptor of purified sewage water: yes; irrigation, receptor of chemical sewage: no

Forest 18
unused 1
number of residents 700,000   inhabitants/km^2 110

Water temperature: min 0·C   max 30°C
Secchi depths: min 20 cm   max 400 cm
Euphotic zone: min 100 cm max 350 cm
O_2 concentration: (open water) min 58%   max 223%
pH 8.3–8.8   Conductivity (μS) 500–600   Alkalinity (mval) 4.4
Average P-conc.   Average N-conc.

Conditions of sediment

North shore high content of org. debris (reed belt). Almost pure silt (90% $CaCO_3$) diminishing towards the S shore, with increasing SiO_2 content. Sand bank near the S shore with 70–90% SiO_2.

Dom. phytoplankton species

Balaton: Nitzschia acicularis, Ceratium hirundinella, Lyngbya limnetica, Cyclotella ocellata, Cyclotella bodanica, Synedra acus v. radians; Balaton NO: Gomphosphaeria lacustris; Balaton SW: Aphanisomenon flos aquae, Anabaena spiroides; An Steinen: Bangia atropurpurea, Cladophora glomerata, Diatoma vulgare

Dom. zooplankton species

Eudiaptomus gracilis, Cyclops spp., Daphnia hyalina, Leptodora kindtii

Dom. macrophytes

Potamogeton perfoliatus, Myriophyllum spicatum, Potamogeton pectinatus Stratiotes aloides, Phragmites australis

Dom. benthic organisms

Chironomids, Oligochaetes, Unionidae, Dreissena polymorpha

Fishes

Abramis brama, Rutilus rutilus, Stizostedion lucioperca, Alburnus alburnus, Pelecus cultratus, Scardinius erythrophthalmus, Acerina cernua, Aspius aspius, Blicca björkna, Silurus glanis.
Some: Cyprinus carpio, Esox lucius, Anguilla anguilla (int.), Hypophthalmichthys molitrix (int.), Lepomis gibbosus (int.)

PHYSICAL AND CHEMICAL MICROSTRATIFICATIONS IN THE SHALLOW LAKE BALATON AND THEIR POSSIBLE BIOTIC AND ABIOTIC ASPECTS

B. ENTZ

Abstract

In the open water of Lake Balaton light is mainly attenuated, during calm weather conditions, by the phytoplankton. Sometimes a temporary temperature stratification can occur during warm, calm days. Light and temperature may indirectly affect the amount of dissolved O_2, pH and redox potential of the water-mass at different depths through their effect on the vegetation. Macrophytes play an important role in this respect, especially in the littoral zone. These effects vary with different species. Two common species, *Potamogeton perfoliatus* and *Myriophyllum spicatum* are the most important ones. Special attention has been drawn to *Ceratophyllum demersum*, because of the peculiarly abrupt vertical changes that occur within a *C. demersum* stand. These conditions are, almost independent of the weather being stable for several weeks. It seems that changes that are normally associated with the water-sediment interface can be studied within dense submerged vegetation but with the advantage of being expanded in time and space. An effect of increasing conductivity associated with *C. demersum* is also reported.

Introduction

Lake Balaton with a surface area of about $600 \, km^2$ is one of the largest lakes of Central Europe. Its length reaches 77 km and its width ranges between 1.5 and 16 km (Fig. 1).

The lake is very shallow, its mean depth being about 3.3 m. The littoral region of the northern wind-protected shore is muddy or sometimes stony, with a wide almost continuous reed belt and, except at its southern end, a rich submerged vegetation.

The southern shore is a wind-exposed, originally sandy beach, with abundant detritus drifts but without any macrovegetation.

The entire lake was, during the first half of the twentieth century, mesotrophic but its trophic level has increased during the last 15 years, particularly in the southern part (Herodek-Tamás, 1975), being nowadays slightly eutrophic at its NE end and extremely hypertrophic in the SW.

This situation affects the underwater light conditions in a conspicuous way (e.g. heavy algal blooms in the SW) causing substantial changes in light penetration into the lake.

It is our main purpose to study the temporary, but sometimes very characteristic, light and heat induced (i.e. sun-energy induced) stratification of the lake together with its physical, chemical and biological aspects.

Methods

The transparency of the water has been measured by the Secchi-disc and relative light intensities have been studied in details by the American Submarine Photometer Model No. 268WA310 of the G.M.MFG & Instrument Corp. Only the total unfiltered light intensities have been measured in the different depths. Besides transparancy values have been determined as measured by a Specord UV VIS spectrophotometer with 4 cm cuvettes at 630 nm wave length.

Temperature (°C) dissolved oxygen ($O_2 \, mg^{-1}$),

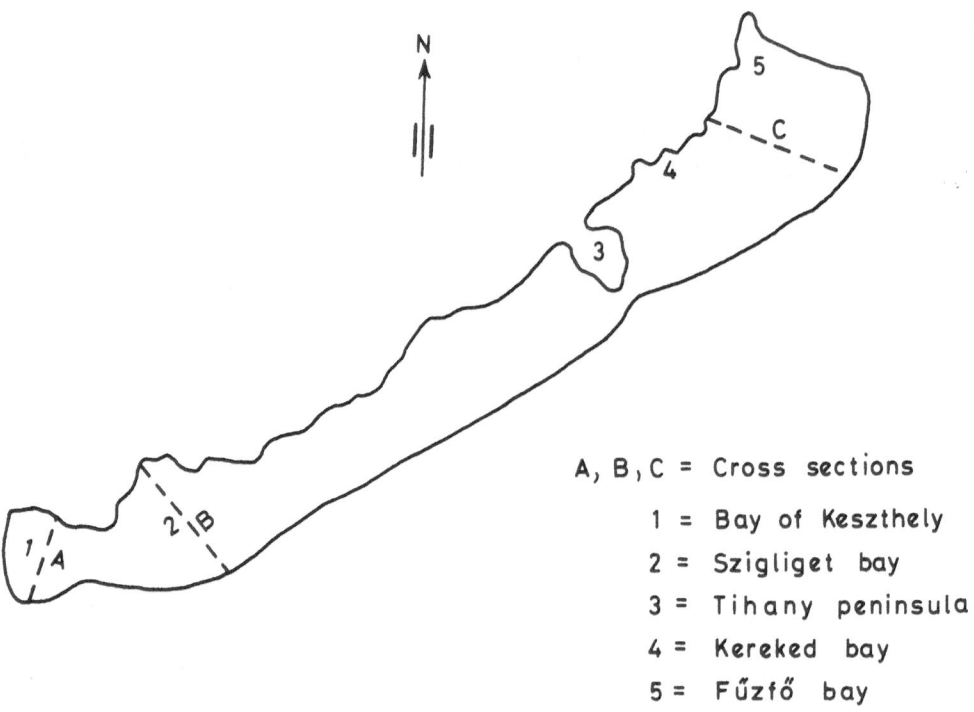

A, B, C = Cross sections
1 = Bay of Keszthely
2 = Szigliget bay
3 = Tihany peninsula
4 = Kereked bay
5 = Fűzfő bay

Fig. 1. Lake Balaton with collecting areas.

electric conductivity ($\mu S = \mu$mhos), pH and redox potential (mV) have been measured simultaneously on the spot by the American Kahlsico – Hydrolab Model 6 surveyor & surface unit.

The stability values have been calculated according to Ruttner (1963).

Results and discussion

Light conditions in the open water

Firstly when wind speeds exceeded 4 m/sec the soft bottom sediments were stirred up all over the lake by wave action, so reducing the transparency of the water almost at once (Entz G. – Sebestyén 1940, Entz, B. in press). This means that the Secchi-transparency was reduced, e.g. from 100 cm to 25 cm, often within a few minutes (Entz–Fillinger, 1961, 1962; Felföldy, 1960; Felföldy–Kalkó, 1958; Tóth–Felföldy–Szabó, 1961).

During calm weather there are remarkable differences in transparency in the different sub-basins of Lake Balaton, being maximal in the NE, medium around the Tihany Peninsula and the central part of the lake, and minimal in the SW (Fig. 2). Comparing the conditions during calm weather firstly for the NE, where 20% of the light that reached the surface could still be recorded at 3.5 m. The same values occurred in the central part of the lake at 0.5 to 1.5 m depths, depending on the weather conditions, while in the SW (Bay of Keszthely) the same light intensities could be found at 0.15 to 0.2 m depths only.

Taking one per cent of the surface light as the lower limit of the photosynthetic zone then this was found usually near the bottom (3.5 m) around the Tihany Peninsula, though in the SW it was in less than 1 m depth (Fig. 2) (Herodek, 1977).

During calm weather the attenuation of light in Lake Balaton gradually increased along a longitudinal section from NE towards SW (Fig. 2), as the increase in the numbers of phytoplanktonic organisms produced a self-shading effect. This is more pronounced in the Bay of Keszthely (Fig. 2) where there are higher algal crops.

The turbidity was greatly increased by heavy storms like that on 4.8.1979, with peak values for

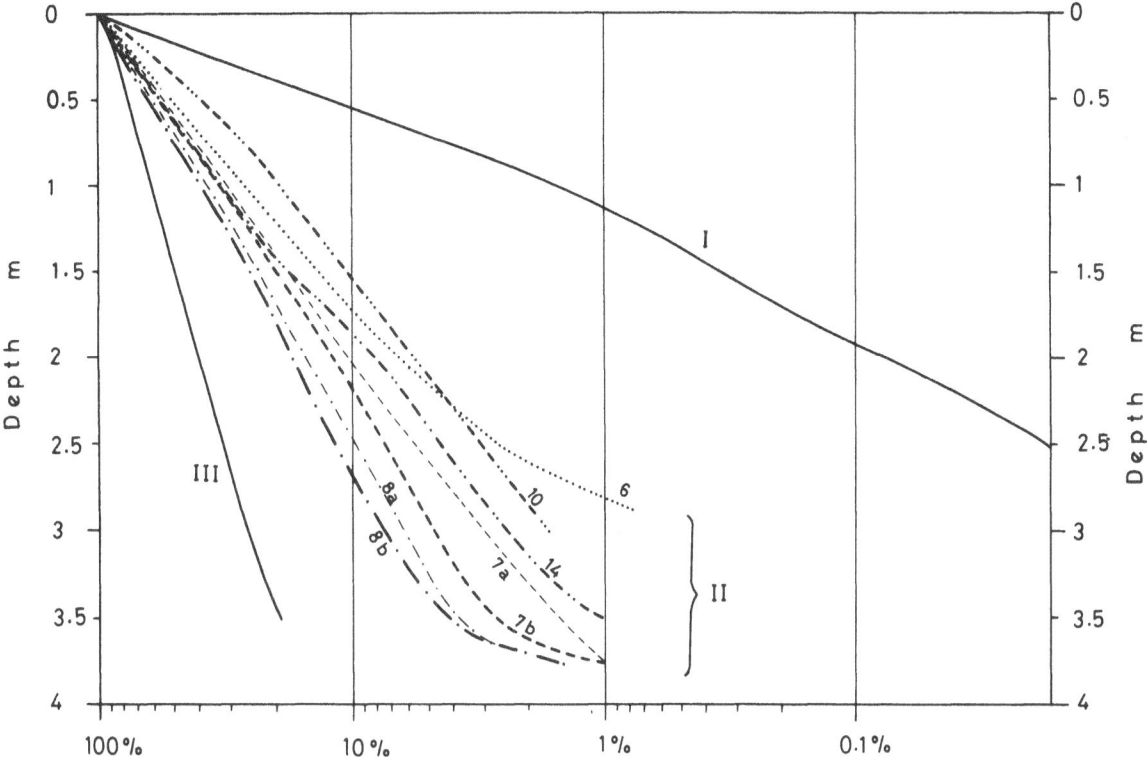

Fig. 2. Relative light penetration in Lake Balaton.
I = Keszthely-bay; II = Different dates at Tihany, 6 to 14 August 1979; III = Füzfö-bay.
N.B.: a = measurements on the boat route, b = measurements outside the boat route.

the wind velocity of 120 km h⁻¹. Only 36 hours after that storm, due to rapid sedimentation, the 1% light level was already at 3 m depth and within a further 24 and 48 hours respectively the bottom was also within the photosynthetic zone (Fig. 2, graphs II, 6, 7a, b, 8a and 8b). Continued recording indicated that another, not very strong wind also reduced the transparancy (Fig. 2, II, 10). These results stressed the fact that decreases as well as increases in transparency may occur quickly in summer. Similar conditions could be concluded from our previous measurements in 1960 (Fig. 3, graphs 25.7 to 31.7).

Similarly in summer 1979 after a heavy storm the amount of suspended matter reached 310 mg l⁻¹ (dry weight), causing a very strong turbidity, but became reduced within two days again to less than 15 mg l⁻¹ (Fig. 4, 13/9). On both these occasions there was bright sunshine.

In contrast in December 1960 when there was absolutely calm but foggy weather there was no remarkable change in transparency of the turbid

water from 14.12.1960 to 17.12.1960 (Fig. 3). Similar results were obtained in January 1979 when after a prolonged and severe storm the lake was frozen on the 4 January. The amount of suspended matter was only slightly reduced under the ice within 16 days from 5.1.1979 to 22.1.1979 i.e. from 30 mg l⁻¹ to 20 mg l⁻¹.

In the meantime the Secchi-transparency only increased from about 15 cm to 35 cm. Also the transparency values, as measured in the spectrophotometer, only increased slightly from 37% (4 January) to 49% (15 January). During this period there was almost no sunshine. Thereafter when the weather became bright both in 1960–61 (Fig. 3, 19.12.60 to 19.1.61) and in 1979 (Fig. 4, 16.1 to 1.2) the transparency increased rapidly.

These coinciding phenomena may stimulate us to puzzle about a possible effect of light (particularly sunshine) on sedimentation. Though a direct correlation seems improbable, an indirect relation cannot be excluded. It may be worthwhile to study these phenomena in more detail.

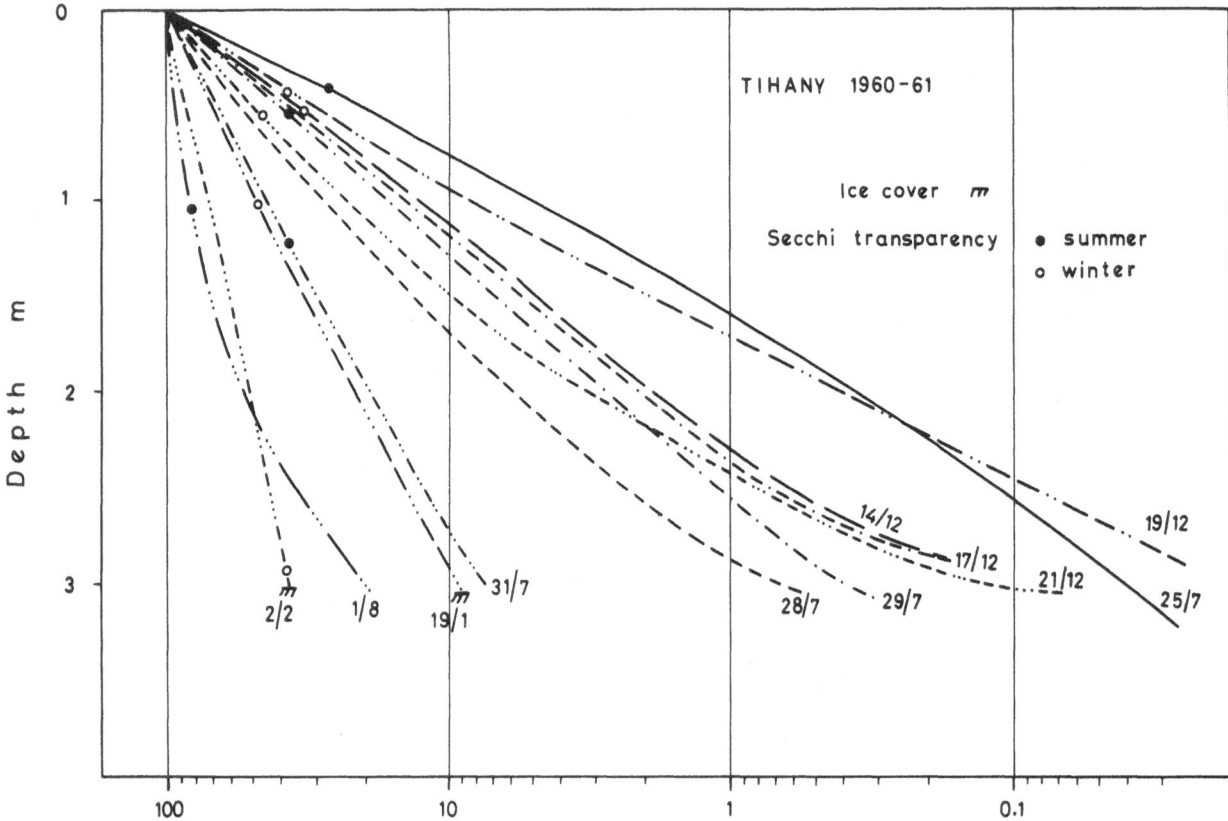

Fig. 3. Relative light penetration to different depths in Lake Balaton in 1960/61.
The numbers 2.2 to 25.7 indicate the dates of measurements.

Aspects of thermal and oxygen stratification in the open water

Even in a shallow lake like Lake Balaton remarkable temperature differences can be formed between the surface and the bottom during long-lasting calm sunny periods.

A good example of this situation occurred on 21.5.1979 (Fig. 5) when, in the area of Szigliget (Fig. 1, Sec. B), the temperature of about three-quarters of the total water-mass surpassed 19°C, while the deeper layers remained as a "cold pool" similar to the period before the warming up process started (14–15°C). The stability of the water mass reached about 2.4 mkg m^{-2} (Ruttner 1963), in spite of its shallowness of 4.1 m. This stability allowed the prolonged existence of this particular stratification (Entz 1979, in press).

In the meantime – as is usual – the high surface temperatures were accompanied by high O$_2$ and pH and low eH values.

	Surface	Bottom
O$_2$ mg 1^{-1}	9.4–10.5	2.2–2.7
pH	8.5	7.8
eH mV	250	280

It is evident that these values are regulated by biological activities. For example the high O$_2$ and pH values are induced by active photosynthesis while the low values near the bottom indicate that within the dark zone no photosynthesis is possible and there is a dominance of catabolic processes, such as O$_2$ consumption of animals, plants and aerobic bacteria.

Similar conditions could be recorded on the same date along the C section (Fig. 1, C), but only in the deepest water layers. Further details about these phenomena are presented in another paper (Entz 1979, in press).

66

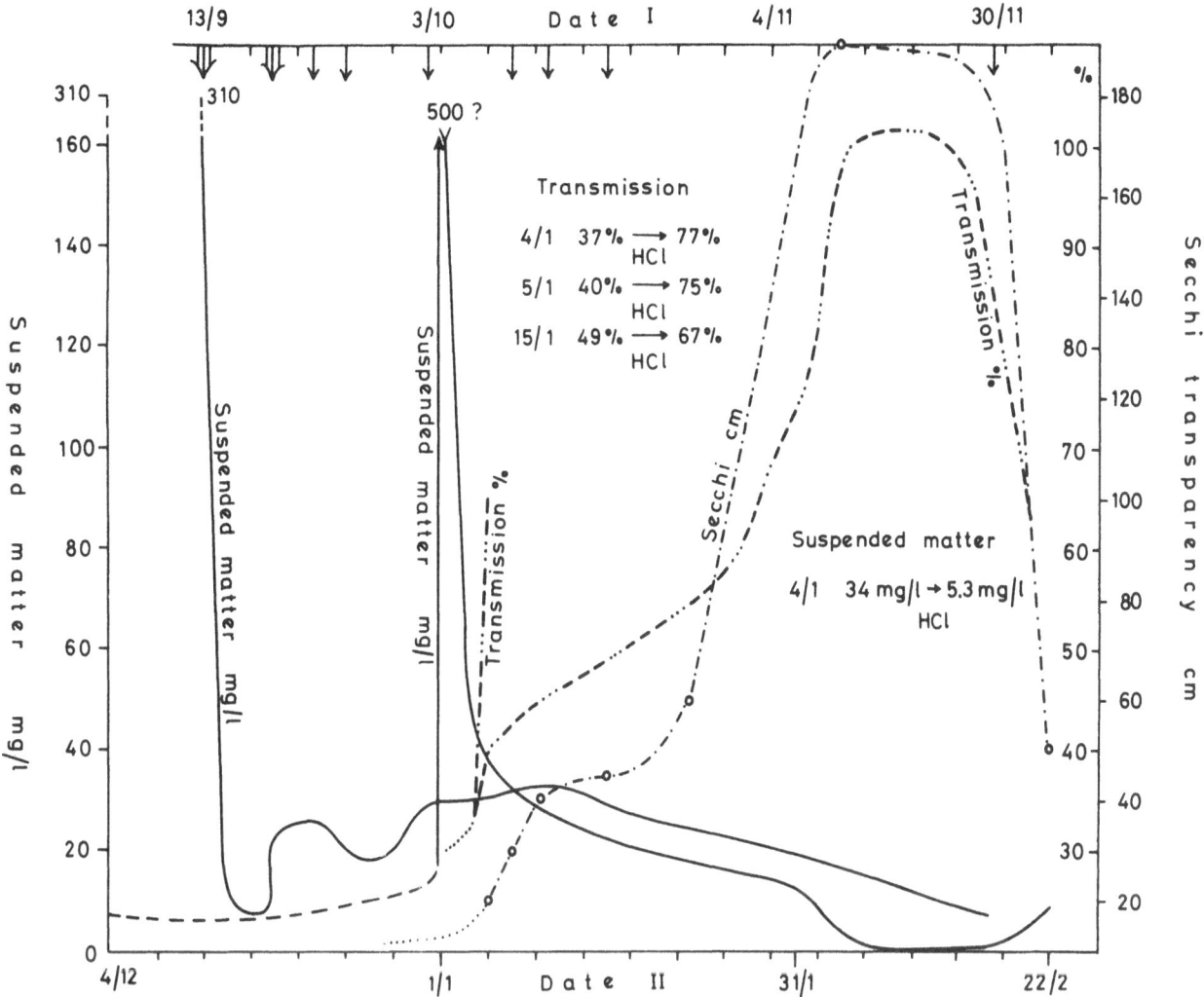

Fig. 4. Light conditions (Transmission in a 4 cm cuvette in per cent — — — : .. — — —; Secchi-transparancy in cm —.—.—.) and the amount of suspended matter in mg l^{-1} (dry weight). All measurements have been carried out in front of Tihany in the open water from 13.9.78 to 30.11.78 (Date I = upper scale), continued from 1.12.78 to 22.2.79 (Data II = lower scale). The curve of the suspended matter starts at 310 mg l^{-1} (upper scale ————), and continues on 1.1.1979 (lower scale ————).

The effects of macrophytes in the littoral zone

In shallow lakes like Lake Balaton there is often a well developed belt of macrophytes in the littoral zone. It is almost always an open question how much importance this belt of macrovegetation plays in the production of organic matter of the lake.

In Lake Balaton the aquatic plants particularly the reed (*Phragmites australis*) and some widespread submerged macrophytes like *Potamogeton perfoliatus* and *Myriophyllum spicatum* have two effects on the production of organic matter. Firstly

as primary producers contributing to the organic biomass and to the formation of organic detritus, the latter playing a very important role in the food web of the lake, and secondly affecting the growth of plankton and benthos, as well as periphytic organisms, in the littoral of the lake through shading effects. This role is especially important as Lake Balaton, being a shallow lake, has a very widespread littoral zone covering, in certain aspects, almost the entire lake.

In our present investigations the effect of macrophytes on the underwater light conditions has been studied (Fig. 6). This figure shows that the

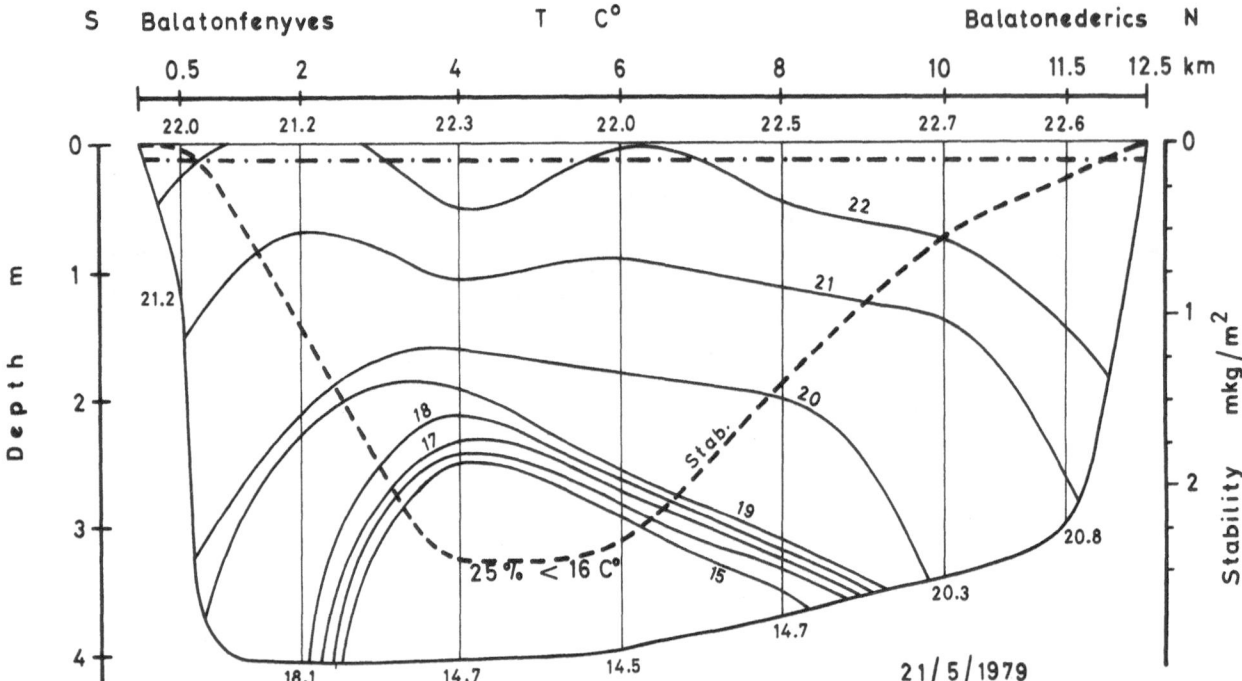

Fig. 5. Temperature conditions (T°C) and stability values (mkg m⁻²) in cross section B between Balatonfenyves and Balatonederics. The vertical lines represent the sampling stations. Ordinate: Depth in m; Abscisse: Distance from the southern shore in km. Stability — — — —.

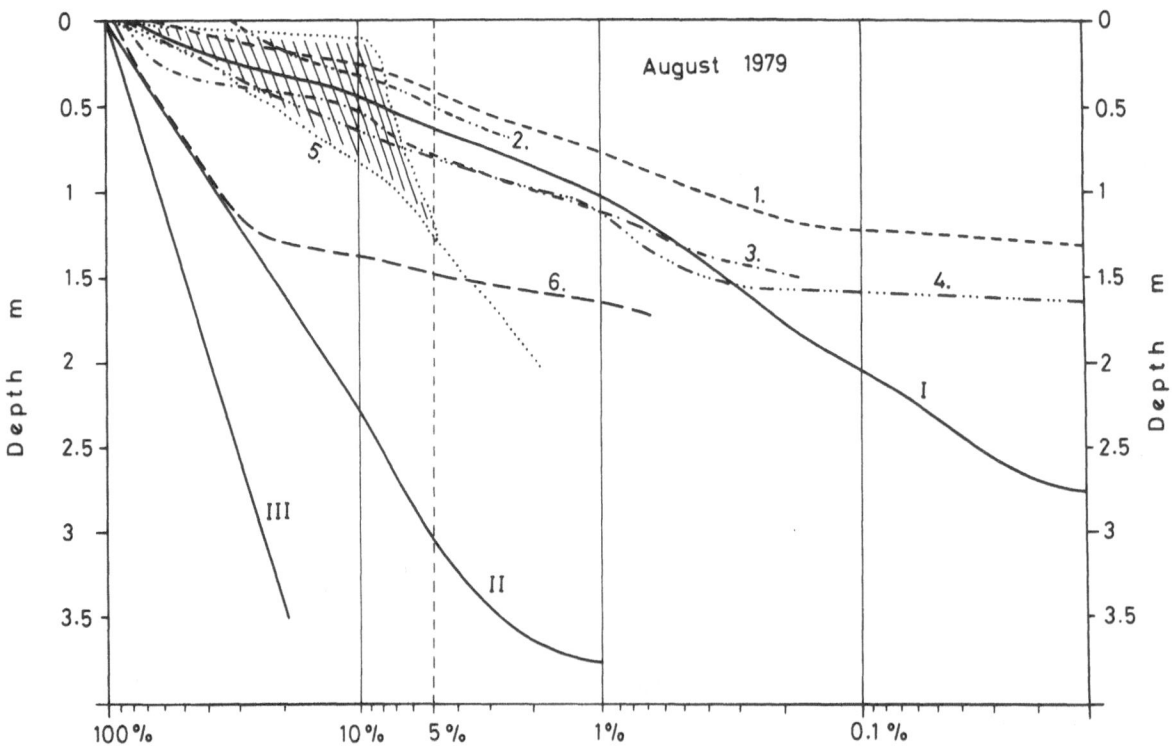

Fig. 6. Relative light penetration in dense stands of different species of macrophytes. I–III Open water conditions. See Fig. 2.
1. Potamogeton crispus - - - -; 2. Phragmites australis - · · - · · -;
3. Stratiotes aloides - · · - · · -; 4. Myriophyllum spicatum - · · · - · · · -;
5. Potamogeton perfoliatus . . . .; 6. Ceratophyllum demersum — — —.

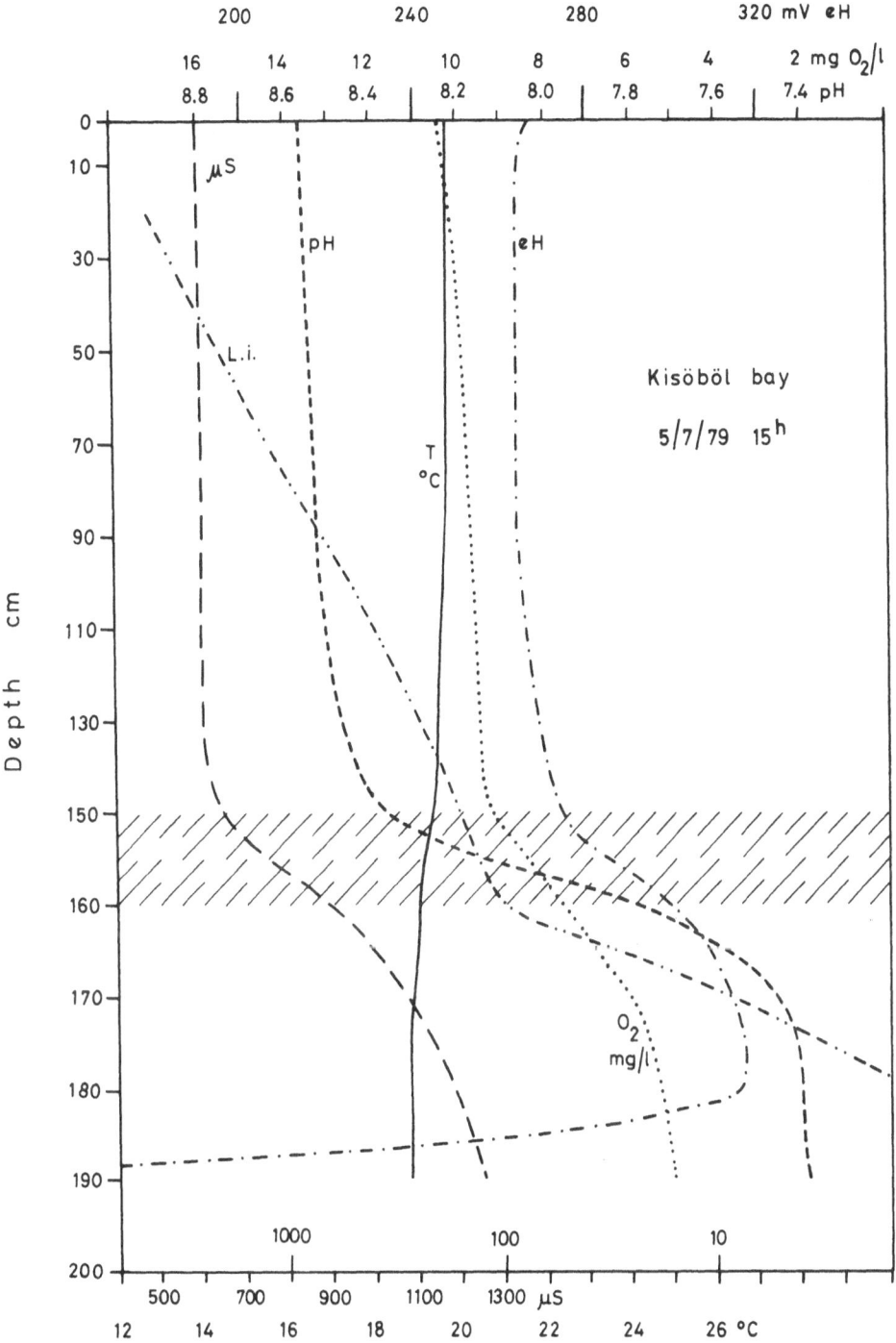

Fig. 7. Conditions in the Kisöböl-bay in a dense stand of Ceratophyllum demersum. Ordinate: depth in cm; abscisse: redox potential (eH) ·–·–·– (200–320), O₂ mg l⁻¹ ······· (16–2), pH ––––––– (8.8–7.4), conductivity μS (= μmhos) — — — — — (500–1300), temperature °C ——————— (12–26). / ₁ / ′ Upper limit of Ceratophyllum carpet

light conditions within dense stands of macrophytes are similar to those in a very rich phytoplankton community (Fig. 6, graphs 1 to 6 and I). This means that within dense carpets of macrophytes light is markedly reduced as compared to conditions in the open water. But the effect is quite different for the different species. A very dense bushy *Potamogeton crispus* stand reduces the light with high efficiency (Fig. 6, gr. 1). Here light is reduced already at 0.5 m depth to 1% of the surface value. Similar to this is the effect of a well developed reed stand. It is remarkable that the shading effect of the emerging part of the reed reduces the light intensity at the surface of the water to about 40% of that of the open water.

At greater depths the underwater conditions are similar to those for *P. crispus. Stratiotes aloides* (Fig. 6, gr. 3) and *Myriophyllum spicatum* have similar but less strong effects. Thus here, the limit of photosynthetic zone (1 per cent of surface light) is at about 1 m depth.

Quite different conditions can be recorded in dense stands of *Potamogeton perfoliatus*. The shoots float like a dense fan on the surface of the

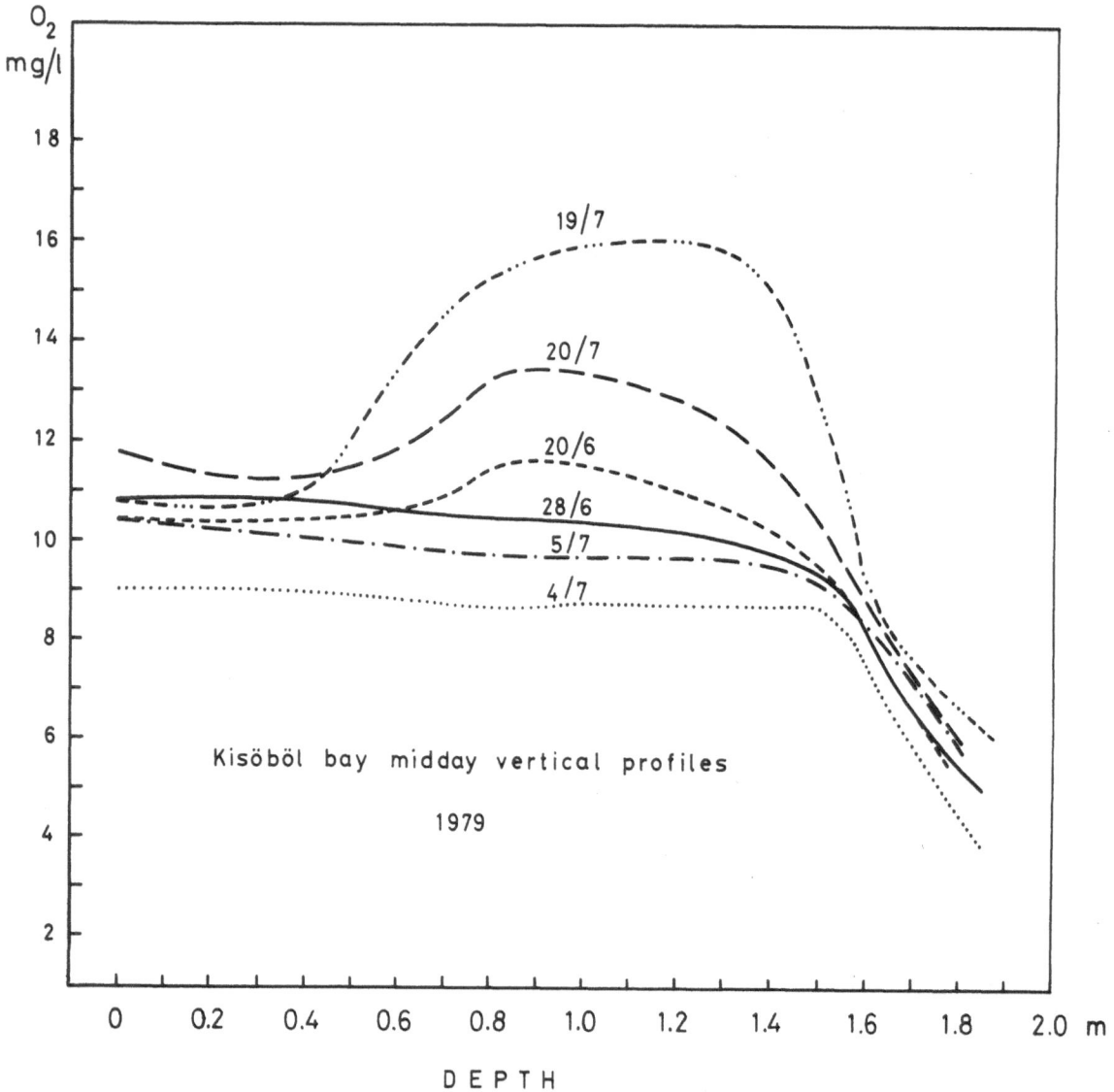

Fig. 8. Changes of oxygen conditions in the Ceratophyllum stand in the Kisöböl-bay from 20 June to 20 July according to depth.

water, while from the bottom to the water surface there is usually only a single stem (Kárpáti–Pomogyi, 1978). As a consequence light is very strongly attenuated to 10–20 per cent at a depth of 10 to 20 cm, where measurements are performed just below the plant shoots. Below the free openings the light reductions are very limited. In greater depths the light conditions are even independent of the place of the measurement (Fig. 6, gr. 5). Light conditions recorded within and above stands of *Ceratophyllum demersum* were quite different. These plants grew on the locality studied to heights of only 30–35 cm above the bottom. But the plants were so crowded that they affected the surrounding water–mass in several ways.

The light above the plants is attenuated only slightly, like in the open water, but on reaching the upper edge of the plant carpet it drops down very sharply (Fig. 6, gr. 6; Fig. 7, L.i. = light intensity). In the meantime there is only a slight change in the temperature, but a remarkable decrease in the O_2 content (from 10 mg l^{-1} to

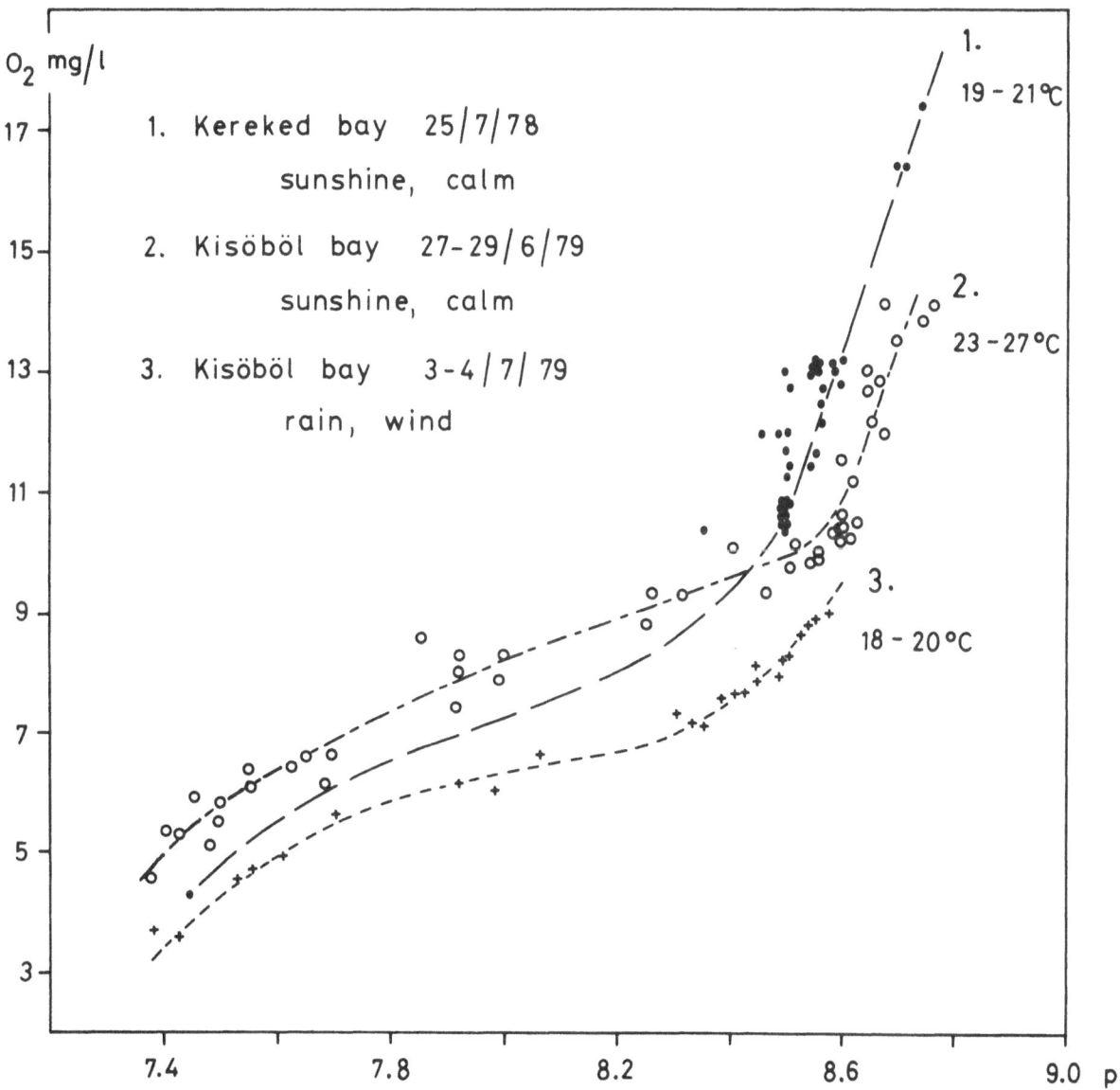

Fig. 9. Relations between dissolved O_2 and pH in different localities by different demperatures.

71

6 mg 1^{-1}) and in pH values (from 8.55 to 7.5). In contrast conductivity was almost doubled (from 580 μS to 1300 μS) and the redox potential first increased and then sharply decreased (Fig. 7). All these phenomena are results of macrovegetation, plankton and bacterial activities. This situation was very stable. Actually these phenomena are similar to the water–mud interface conditions but have the advantage for investigations that the depth intervals are expanded to a layer of 20 to 40 cm – and this remains stable for several weeks at the same spot (Fillos, 1977; Viner, 1975).

In June to August 1979 the same or at least very similar conditions could be observed within the thick stand of *Ceratophyllum* though in the meantime, above the weed-stand, in the free waterlayer, the influences of the ambient air conditions (e.g. rain or sunshine, windy or calm days, hot or cold temperature) were very remarkable (Fig. 8). This underlines the high stability of the habitats within the dense weed stands.

Finally, looking at the dissolved oxygen and the pH values obtained simultaneously but taken in different biotopes under different meterological conditions, a definite trend could be assumed (Fig. 9). Higher oxygen values could be related to identical pH values by higher water temperatures than by lower ones. Accordingly the curves showing this relation change neither according to the hour of the day, nor to locality but are different according to the water temperature.

These conditions and a pronounced bend within the curves around pH 8.6 allow us to suggest a certain effect of temperature on the dissolved carbonic acid – bicarbonate – carbonate system in Lake Balaton, resulting in higher turbidity of the lake-water in the summer than in the winter season under otherwise similar conditions.

Because the mechanism of this system is so far not clearly understood, further detailed studies are required in this field.

References

Entz, B. 1980. Microstratification phenomena of Lake Balaton, a shallow lake in Central Europe. Hung. Acad. Press (in press).

Entz, B. & E. Fillinger, M. 1961. Angaben zur Kenntnis des Lichtklimas des Balaton (Über die Ursache der Wassertrübungen und deren Auswirkungen). Annal. Biol. Tihany 28: 49–89.

Entz, B. & E. Fillinger, M. 1962. Angaben zur Kenntnis des Lichtklimas des Balaton II. Über die Lichtverhältnisse im Wasser des Zugefrorenen und schneebedeckten Sees. Annal. Biol. Tihany 29: 65–74.

Entz, G. & Sebestyén, O. 1940. Das Leben des Balatonsees. Arb. Ung. Biol. Forsch. Inst. Tihany 12: 1–169.

Felföldy, L. 1960. Apparent photosynthesis of *Potamogeton perfoliatus* L. in different depths of Lake Balaton. Annal. Biol. Tihany 27: 201–208.

Felföldy, & Kalkó, Zs. 1958. The rate of photosynthesis and underwater radiation in Lake Balaton. Observations of summer 1957. Annal. Biol. Tihany 25: 302–329.

Fillos, J. 1977. Effect of sediments on the quality of the overlying water. In: Interactions between sediments and fresh water, Golterman H. L. (ed.) Junk Publ. The Hague 1977, 266–271.

Herodek, S. 1977. Recent results of phytoplankton research in Lake Balaton. Annal. Biol. Tihany 44: 181–198.

Herodek, S. & Tamás, G. 1975. The primary production of phytoplankton in the Keszthely-basin of Lake Balaton in 1973–1974. Annal. Biol. Tihany 42: 175–190.

Kárpáti, V. & Pomogyi, P. 1978. Accumulation and release of nutrients by aquatic macrophytes. Symp. Biol. Hung. 19: 33–42.

Ruttner, F. 1963. Fundamentals of Limnology. Toronto, University of Toronto Press, pp. 295.

Tóth, L., Felföldy, L. & Szabó, E. 1961. On some problems of production measurements in Phragmiteta in Lake Balaton. Annal. Biol. Tihany 28: 169–178.

Viner, A. B. 1975. The sediments of Lake George (Uganda) 3. The uptake of phosphate. Arch. Hydrobiol. 76. 393–410.

CHANGES IN THE STRUCTURE OF PHYTOPLANKTON IN LAKE BALATON AS A RESULT OF EUTROPHICATION

L. VÖRÖS & J. NÉMETH

Abstract

In the shallow (mean depth 3.14 m) and elongate Lake Balaton, the horizontal distribution of phytoplankton biomass shows a gradual increase from east to west related to the changes of plant nutrients. In the more eutrophicated areas of the lake the values of diversity were higher than in the less eutrophicated ones, but good parallelisms between diversity and biomass were not usually detected. With the aid of cluster analysis three water quality areas were distinguishable in the lake on the basis of phytoplankton patterns but the position of their border lines changed with time.

Introduction

The main water quality problem of Lake Balaton is due to eutrophication. This process is observable everywhere in the lake, but the different areas are not equally eutrophicated (Herodek, 1977). The most polluted part of the lake is the western-basin where the greatest chlorophyll-a content that has been measured is more than $100\ \mu g\,l^{-1}$. The Zala river provides 30–60% of surface of the run-off flowing into Lake Balaton; it also provides approximately 20–50% of phosphorus and the 30–40% of nitrogen transported by the water streams flowing into the western-basin. The less polluted part is the eastern-basin where the highest chlorophyll-a content is not more than $20\ \mu g\,l^{-1}$.

The quantities of plant nutrients are largest in the western-basin with a gradual decrease east-wards (Tóth, L., 1976). A similar pattern is shown to the chlorophyll-a (Tóth, L., 1976; Tóth, F. et al., 1976), and the biomass of algae (Tamás, 1975; Vörös, 1979), the primary production (Herodek, 1977) and the bacterial counts (Oláh, 1973).

The present paper deals with the qualitative and quantitative phytoplankton data of years 1978 and 1979 in order to establish:

- the horizontal distribution of phytoplankton biomass
- the species diversity of the phytoplankton in the different areas
  the different water quality areas on the basis of patterns of phytoplankton distribution using several cluster analysis methods.

Material and methods

The samples were collected from 10 stations located along the long axis of the lake (Fig. 1) in April, August and September 1978 and in January 1979. The last samples (January 1979) were taken below the ice cover.

The samples derived from 4 different depths were mixed to represent the whole water column and were fixed by Lugol's solution containing acetate. Samples were counted by Utermöhl's technique, and the mass of phytoplankton was calculated from the cell volumes. For calculation of the diversity and similarity we used the biomass data instead of individual numbers because the volume

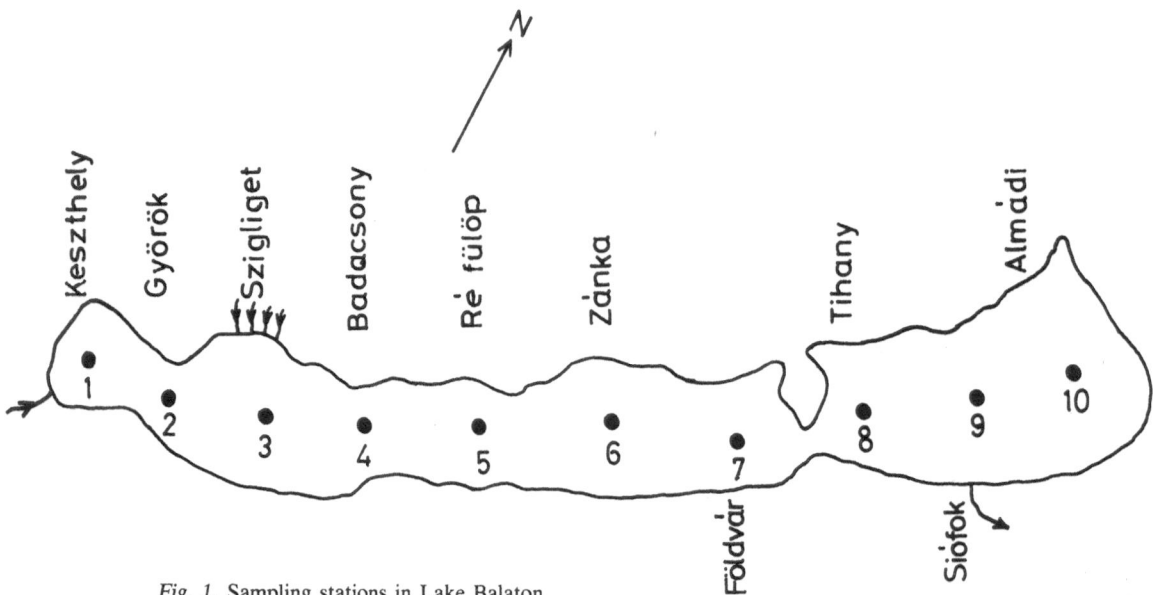

Fig. 1. Sampling stations in Lake Balaton

of individual cells of different species was very different (Hallegraeff & Ringelberg 1978), there was often a 4 to 5 order of magnitude difference between cell volumes.

The diversity index and equitability were calculated according to the following equations (Shannon & Weaver 1963; Pielou 1975):

$$H'' = - \sum_{i=1}^{s} \frac{n_i}{N} \log_2 \frac{n_i}{N}$$

$$J'' = \frac{H''}{\log_2 s}$$

Where H'' is the diversity,

s is the number of species occurring in the sample

n_i is the biomass belonging to the ith species,

N is the total biomass

J'' is the equitability

For cluster analysis the average linkage method (UPGMA) was chosen (Sneath and Sokal, 1973) and applied to several similarity measures according to Czekanowsky (1909), Kulczynski (1928) and Ruzička (1958).

$$S_{\text{Czekanowsky}} = \frac{2 \sum_{i=1}^{i=s} \min (n_{1i}, n_{2i})}{N_1 + N_2}$$

$$S_{\text{Kulczynski}} = \frac{1}{2} \left[\frac{1}{N_1} + \frac{1}{N_2} \right] \sum_{i=1}^{i=s} \min (n_{1i}, n_{2i})$$

$$S_{\text{Ruzička}} = \frac{\sum_{i=1}^{i=s} \min (n_{1i}, n_{2i})}{N_1 + N_2 - \sum_{1=1}^{i=s} \min (n_{1i}, n_{2i})}$$

Where S is the similarity measure,
min (n_{1i}, n_{2i}) is the minimum value of the biomass belonging to the i-th species in the pair of samples compared,
N_1 and N_2 are the total biomass of sample 1 and 2 respectively, s is the number of species in the pair of samples compared.

Results and discussion

Biomass

In April 1978 (Fig. 2) the horizontal distribution of biomass was extraordinary because it was highest in the middle of the lake. This phenomenon which had not been observed previously, was probably caused by the different seasonal succession in the different localities. In August and September 1978 and January 1979 (Figs. 3–5) the distribution of algal biomass was typical, the highest value was measured in the Keszthely-basin (station No 1) and decreased more or less gradually along the long axis of the Lake Balaton in accordance with earlier results (Tamás 1975;

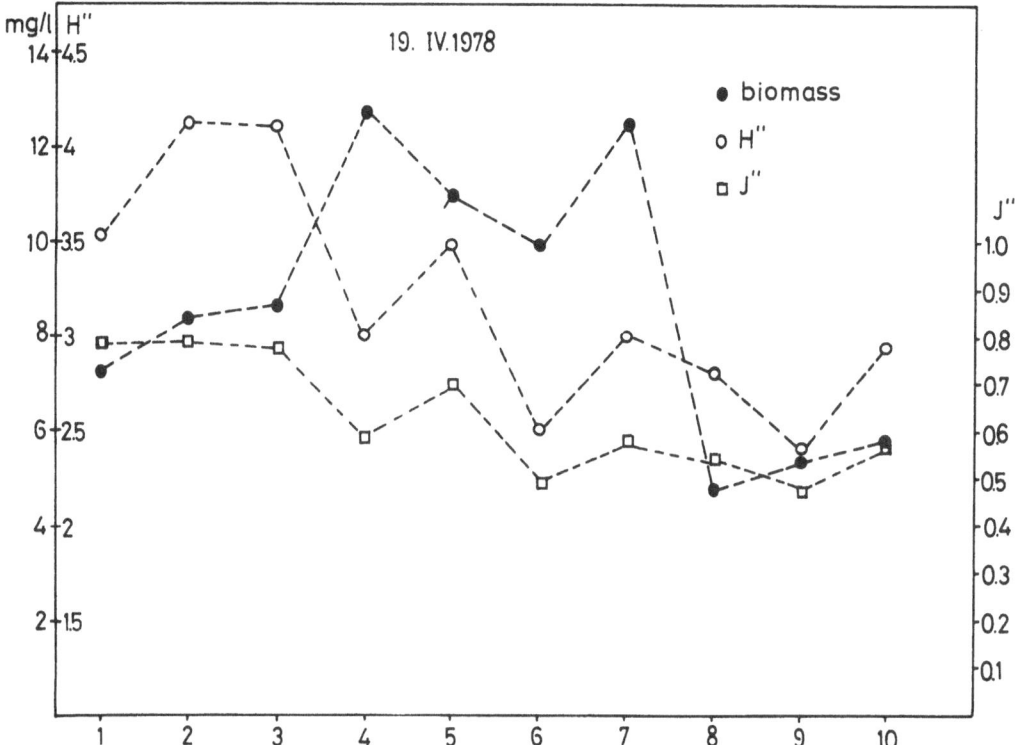

Fig. 2. The horizontal distribution of phytoplankton biomass and the changes of diversity (H'') and equitability (J'') in April 19, 1978

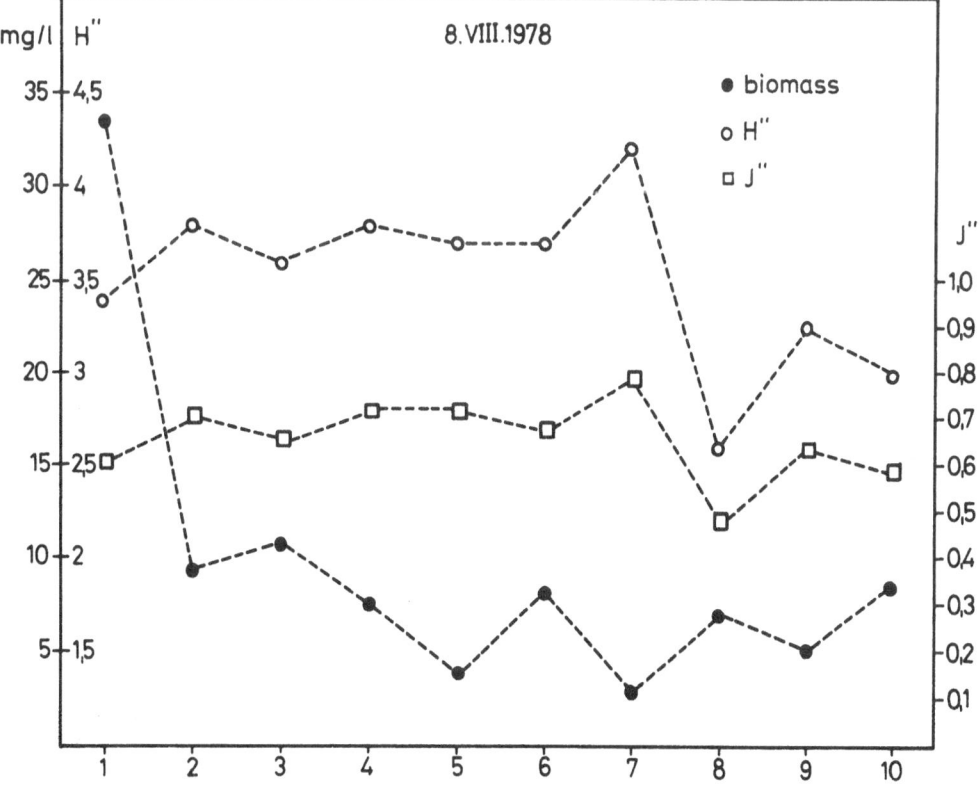

Fig. 3. The horizontal distribution of phytoplankton biomass and the changes of diversity (H'') and equitability (J'') in August 8, 1978

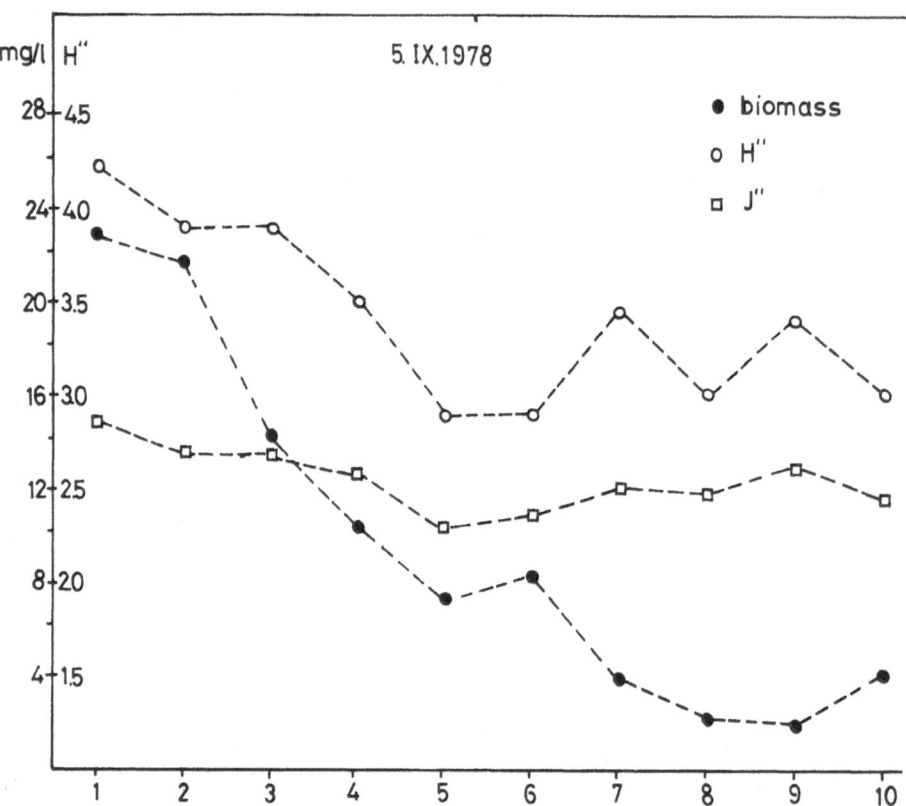

Fig. 4. The horizontal distribution of phytoplankton biomass and the changes of diversity (*H''*) and equitabil

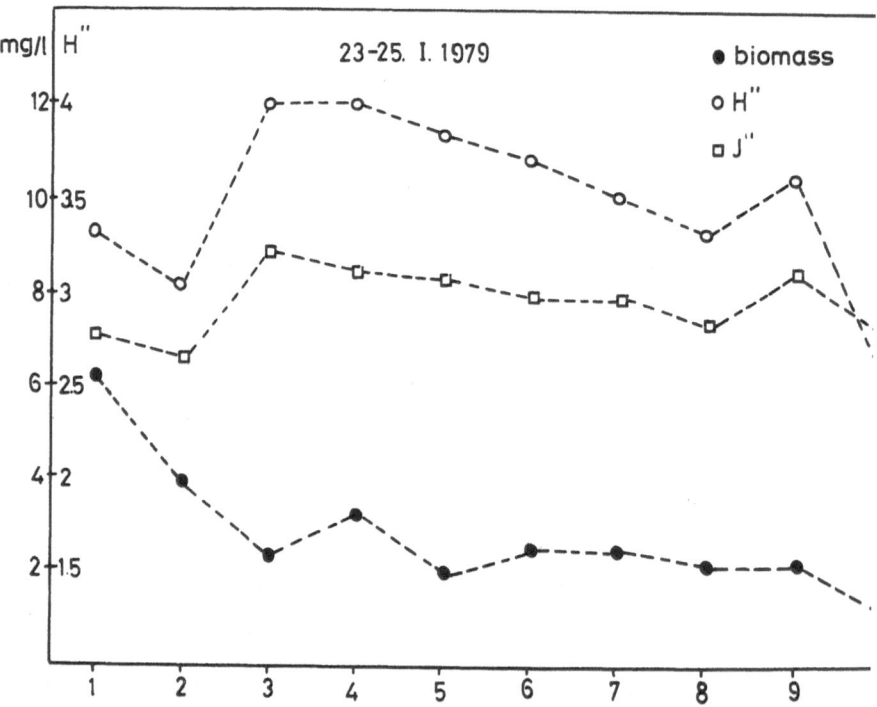

Fig. 5. The horizontal distribution of phytoplankton biomass and the changes of diversity (*H''*) and equitability

Vörös 1979). The summer maximum of biomass in the Keszthely-basin (34 mg l^{-1}) indicates the rapid eutrophication because in the 1960s it was not more than 8 mg l^{-1}.

In the lake the Chlorophyta had the greatest number of species followed by Bacillariophyta and Cyanophyta. The number of species was greatest in summer and lowest in winter. In spring the middle part of the lake and the eastern-basin were dominated by *Cyclotella bodanica* Eulenst. the western-basin by *Stephanodiscus hantzschii* var. *pusillus* Grun. In summer in the less eutrophicated area *Gomphosphaeria lacustris* Chodat, *Ceratium hirundinella* (O. F. Müll.) Schrank, and *Botryococcus braunii* Kütz. prevailed while in the western part of the lake the blue-greens, *Aphanizomenon flos-aquae* (L.) Ralfs, *Aphanizomenon issatschenkoi* (Uss.) Proschkina- Lavrenko, *Anabaena spiroides* Kleb., and *Anabaena aphanizomenoides* Forti. In winter in the whole lake *Cryptomonas erosa* Ehr. dominated, its biomass increasing monotonously from east to west.

Diversity

Seasonal changes of diversity were not observable on the basis of these four investigations, but the earlier more frequent measurements demonstrated a definite annual cycle in Lake Balaton (Vörös, 1979).

The lowest value of diversity index was observed in the eastern basin of the lake together with the lowest trophic state. In the more eutrophicated areas (the middle part of the lake and the western basin) the values of diversity were higher than in the less eutrophicated ones, but good parallelism between diversity and biomass was not usually detected (Figs. 2–5).

In the case of Lake Balaton the increasing eutrophication was not followed by a decrease in diversity. The eutrophication of the lake has not yet attained the level above which further elevation of the biomass is paralleled by a decrease in the number of species and diversity in the phytoplankton as frequently observed in hypertrophic waters.

Spatial pattern of phytoplankton

In April 1978 according to the dendrograms (Fig. 6) the sampling sites were separated into 3

distinct clusters (1–2, 3–7, 8–10). The different similarity measures used in this study always grouped the stations in the same manner but at the different level of similarity, therefore we publish the dendrograms only on the basis of the Czekanowsky measure.

In August 1978 (Fig. 6) three parts of the lake were also distinguishable (1, 2–8, 9–10). The Keszthely-basin was very different from all other sampling sites. At this time the similarity between the stations was usually low and the clustering did not give such clear groups as before.

In September 1978 (Fig. 7) the dendrograms again showed three clearly distinguishable groups (1–3, 4–6, 7–10). The level of similarity between sampling sites was higher than in August.

In January 1979 (Fig. 7) the sampling sites were also separated into three groups (1–2, 3–8, 9–10) and the similarity between the stations was highest of all the cases. On the basis of these results there was a seasonal change in the level of similarity, it was higher in winter and spring than in summer and autumn.

Summarizing the results of the cluster analysis it is possible to distinguish three water quality areas in Lake Balaton on the basis of phytoplankton patterns, but their border lines change with time (Fig. 8). These changes are probably closely related with the circulation pattern of the lake (Györke, 1975).

Finally we think that the change of the border lines has great importance in the planning of future research of Lake Balaton. When examining chemical or biological phenomena it will be necessary to work at localities far away from the moving border lines.

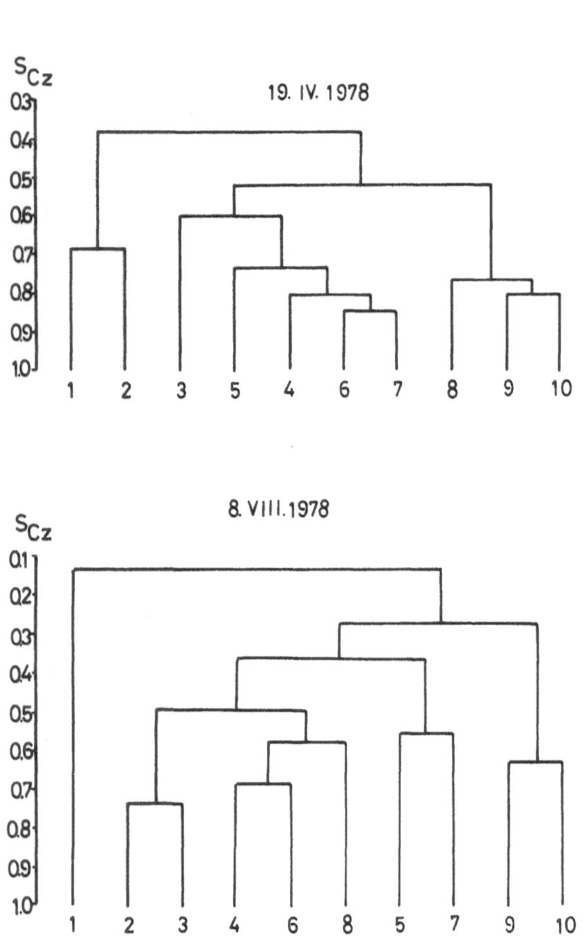

S_{Cz}

19. IV. 1978

0.3
0.4
0.5
0.6
0.7
0.8
0.9
1.0

1 2 3 5 4 6 7 8 9 10

S_{Cz}

5. IX. 1978

0.2
0.3
0.4
0.5
0.6
0.7
0.8
0.9
1.0

1 2 3 4 5 6 7 8 9 10

S_{Cz}

8. VIII. 1978

0.1
0.2
0.3
0.4
0.5
0.6
0.7
0.8
0.9
1.0

1 2 3 4 6 8 5 7 9 10

S_{Cz}

23–25. I. 1979

0.4
0.5
0.6
0.7
0.8
0.9
1.0

1 2 3 5 4 6 7 8 9 10

Fig. 6. The dendrograms of samples in April 19, 1978 and in August 8, 1978

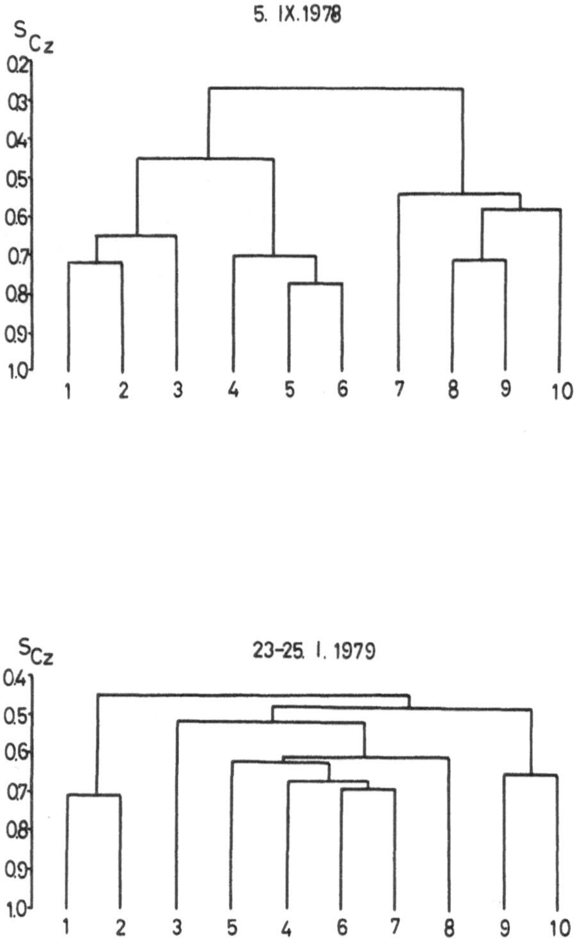

Fig. 7. The dendrograms of samples in September 5, 1978 and in January 23–25, 1979

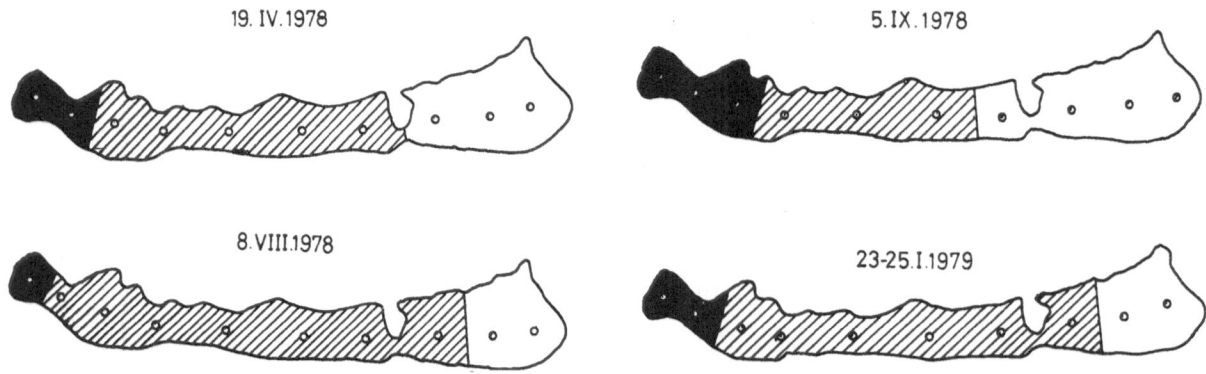

19. IV. 1978

5. IX. 1978

8. VIII. 1978

23-25. I. 1979

Fig. 8. The temporal changes of the different water quality areas.

78

References

Czekanowski, J. 1909. Zur Differentialdiagnose der Neandertalgruppe. Korrespbl. dt. Anthrop. Ges. 40: 44–47.

Györke, O. 1975. Water and sediment motion in the southeastern area of Lake Balaton, characterizing bed formation. VIZDOK (in Hungarian).

Hallegraeff, G. M. & Ringelberg, J. 1978. Characterization of species diversity of phytoplankton assemblages by dominance-diversity curves. Verh. Internat. Verein. Limnol. 20: 939–949.

Herodek, S. 1977. Recent results of phytoplankton research in Lake Balaton. Annal. Biol. Tihany 44: 181–198.

Kulczynski, S. 1928. Die Pflanzenassotiationen der Pieninen. Bull. Int. Acad. Pol. Sci. Lett. Cl. Sci. Math. Nath. Ser. B. 1927, Suppl. 2, 57–203.

Oláh, J. 1973. The biomass and production of bacterioplankton in Lake Balaton. Hidrológiai Közlöny 53: 348–356. (in Hungarian)

Pielou, E. C. 1975. Ecological Diversity. Wiley – Interscience, 165 p.

Ruzička, M. 1958. Anwendung matematisch-statistischer Methoden in der Geobotanik. (Syntetische Bearbeitung von Aufnahmen.) Biologie (Bratislava) 13: 647–661.

Sneath, P. H. A. & Sokal, R. R. 1973. Numerical Taxonomy. Freeman and Co., 573 p.

Shannon, C. E. & Weaver, W. 1963. The mathematical theory of communication. University of Illinois Press, Urbana.

Tamás, G. 1974. The biomass changes of phytoplankton in Lake Balaton during the 1960s. Annal. Biol. Tihany 41: 323–342.

Tamás, G. 1975. Horizontally occurring quantitative phytoplankton investigations in Lake Balaton, 1974. Annal. Biol. Tihany 42: 219–279.

Tóth, F., Vörös, L. & Németh, J. 1976. Investigations of chlorophyll-a and the phytoplankton of Lake Balaton. Balatoni Ankét 30 September–1 October, 1976, pp. 1–12. (in Hungarian)

Tóth, L. 1976. Water quality studies of Lake Balaton. Annual Hydrological Information Series on Lake Hidrology, ed. Baranyi, S. (in Hungarian)

Vörös, L. 1979. Phytoplankton biomass and algal counts in Lake Balaton. Balatoni Ankét May 17–18, 1979. III.A, 16: 1–16. (in Hungarian)

Vörös, L. 1979. The phytoplankton of Lake Balaton in 1976. Botanikai Közlemények (in press). (in Hungarian)

ABOUT FEEDING CONDITIONS OF BREAM, (*ABRAMIS BRAMA* L.) IN LAKE BALATON

I. TÁTRAI

Abstract

The quantitative composition of the chironomid population, as the main food resource for adult bream, is described in relation to changes in numbers and biomass during the year. Adult bream feed mainly on the bottom fauna and to a smaller extent on zooplankton. Bream consumed, during the period of study, about 30 to 35% of the standing crop of Chironomidae. Zooplankton were less affected as adult bream consumed only 4 to 6% of the biomass of zooplankton.

Introduction

The majority of fish production in Lake Balaton consists of bream. The production of age-groups 3+ to 7+, which are the most intensively exploited groups by the commercial fishery, amounts to 46.7 kg ha^{-1} (Bíró & Garádi, 1974). As a consequence of the breams large population among the non-piscivorous fish it plays an important role in utilizing the benthos, mainly the larvae of chironomids.

Some data on the feeding of bream in Lake Balaton have been published by Entz & Lukacsovics (1975), particularly during the winter months, when feeding was intensive and the food consisted mainly of benthic organisms.

Our present knowledge concerning the bream's food and feeding rate in Lake Balaton during the summer is still insufficient.

The aim of our studies was to determine the food resources of the bottom zone, especially regarding the larvae of chironomids, that are available for bream. These were based on analyses of the contents of the alimentary canal, estimates of food composition, and the rate of feeding of bream in Lake Balaton, as well as analyses of bottom samples.

Area, material and methods

Lake Balaton has an area of 593 km^2 and its mean depth reaches 3.14 m. Its trophic state at Tihany in the Szigliget basin near Badacsony can be taken as intermediate between eutrophic and hypertrophic levels, and in the Keszthely basin as strongly hypertrophic (Herodek, 1977).

The benthic material was collected in 1977 from May until November at monthly intervals. Samples were taken at 3 sites between Tihany and Siófok, between Badacsony and Fonyód, and in the Keszthely-bay in the open water. On each station 5 samples were taken with an Ekman-Birge mud-sampler, catching area 225 cm^2. After washing the material on a sieve (21 × 30 cm, mesh size 0.5 × 0.5 mm) the samples were put into vessels with distilled water in order to empty the alimentary canals. The fresh weight of the larvae was assessed by weighing them on analytical balance after drying on blotting paper.

The material was taken from fish, caught by the Balaton Fishing Company, Siófok, at stations

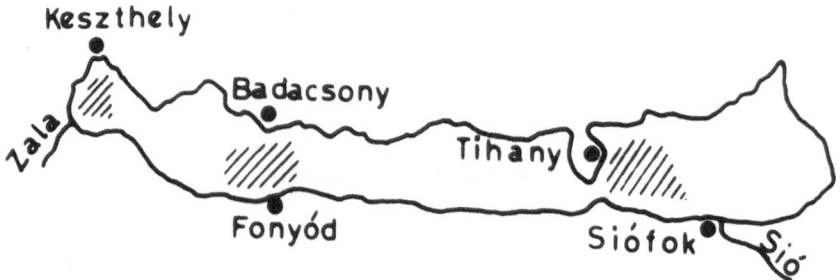

Fig. 1. Stations in Lake Balaton.

Tihany-Siófok, Badacsony-Fonyód and Keszthely-bay (Fig. 1). Mature specimens of age-groups 3+ to 6+ were examined. The fish were immediately weighed, and the alimentary tracts preserved in 5% formalin. Altogether 450 alimentary tracts were studied. Each was weighed separately and examined using a stereo-microscope. The food animals or their remains were counted, measured, identified and their original weights recalculated

(Sebestyén, 1958; Nauwerck, 1963). The proportions of plant material and detritus content were calculated assuming that 1 ml corresponds to 1 g fresh weight (Hunt, 1960).

Results and discussion

The numbers of chironomids, as the main food resource of bream, fluctuated during the sampling

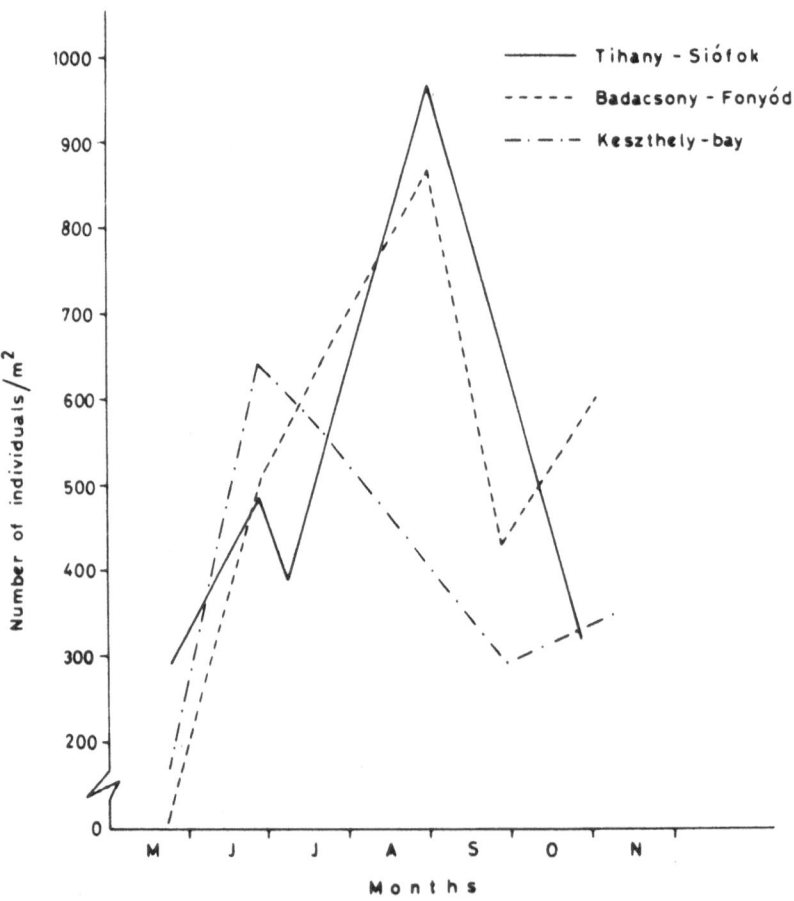

Fig. 2. Dynamics of the numbers of chironomids in three basins of Lake Balaton in 1977.

period (Fig. 2) attaining a maximum of 980 ind. m^{-2} at Tihany. The mean number of chironomids was 491 at Tihany, 503 at Dacsony and 402 ind. m^{-2} in the Keszthely-bay respectively. These differences in numbers were due to the fact that the larvae of the small forms of the subfamilies Orthocladiinae and Tanypodinae were the most numerically abundant of the Chironomidae at Tihany and at Badacsony. The number of Chironomidae depends similarly on the size of the particles in the mud (Entz, 1965). In the Keszthely-bay large forms – chiefly *Chironomus plumosus* – were dominant. There was good agreement with numbers and spatial distribution of chironomid larvae in 1976 (Tátrai, 1977). According to the first quantitative analysis (Entz, 1954), 60% of the zoobenthos comprises chironomid larvae in Lake Balaton.

The dynamics of chironomid biomass shows seasonal changes-high values in the late spring period (June, 4–5 g m^{-2}) an obvious decrease in summer (2–4 g m^{-2}) and again an increase in au-tumn (September, 4–8 g m^{-2}) followed by a decrease at the end of October (1.5–3 g m^{-2}, Fig. 3). Wojcik (1970) and Kajak & Dusoge (1973) have observed similar dynamics of benthic chironomid biomass in Lake Warniak, Poland. Similar cycles were observed in other environments, too (Wojcik-Migala, 1965 – in fish ponds; Kajak, 1960 – in old river beds). During 1964 *Chironomus plumosus* reached a maximum of 1900 ind. m^{-2} and 32 g m^{-2} fresh weight in the middle of the lake in autumn (Entz, 1965). In 1973–74 *Tanypus punctipennis* reached a maximum of 9800 ind. m^{-2} (35 g m^{-2}, 1.4 g C m^{-2}) in February. In summer the biomass dropped to 2.5 g m^{-2} (0.1 g C m^{-2}, Oláh, 1976). It is evident that biomass as well as species composition of the zoobenthos is affected by the increasing eutrophication of Lake Balaton.

The biomass of planktonic crustaceans, as accidental but non-essential food for bream, differs in the regions mentioned in Lake Balaton. During the summer it reaches 1.09 g m^{-3} wet weight within

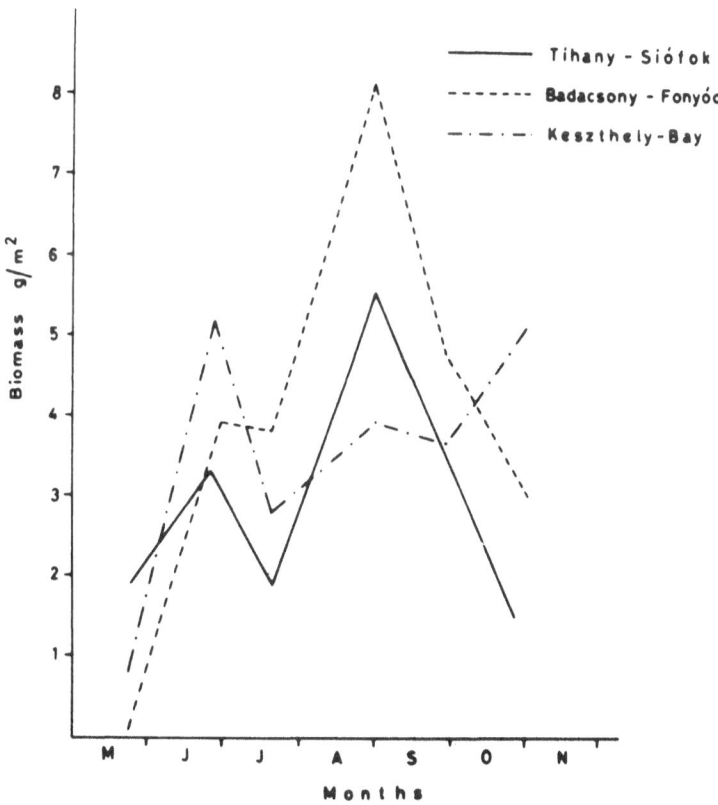

Fig. 3. Dynamics of the biomass of chironomids in three basins of Lake Balaton in 1977.

the Keszthely-bay, 0.91 g m^{-3} in the basin of Szigliget and 0.66–0.75 g m^{-3} in other areas (Ponyi, 1975). These quantities are low as compared to these in other eutrophic lakes (Hillbricht-Ilkowska et al., 1973).

Animal food was 95% of the wet weight of the food of bream, with body weights ranging from 125 to 434 g, during the period investigated. Plant food was found in 56% of the alimentary tracts analyzed and reached 5% of all food weight.

A marked dominance of the chironomid larvae was observed in the animal food of bream in all the three basins of the lake (Fig. 4). They were the most numerous and the most frequent food components varying from 60 to 80% of food weight. The others, Oligochaeta, Ephemeroptera, Odonata, Mollusca, detritus and algae, were found in all of the intestines examined and ranged from 10 to 40% of the food weight.

The planktonic crustaceans play an insignificant role in the rate of feeding of bream. Consumption rate of crustaceans did not exceed 8% of the

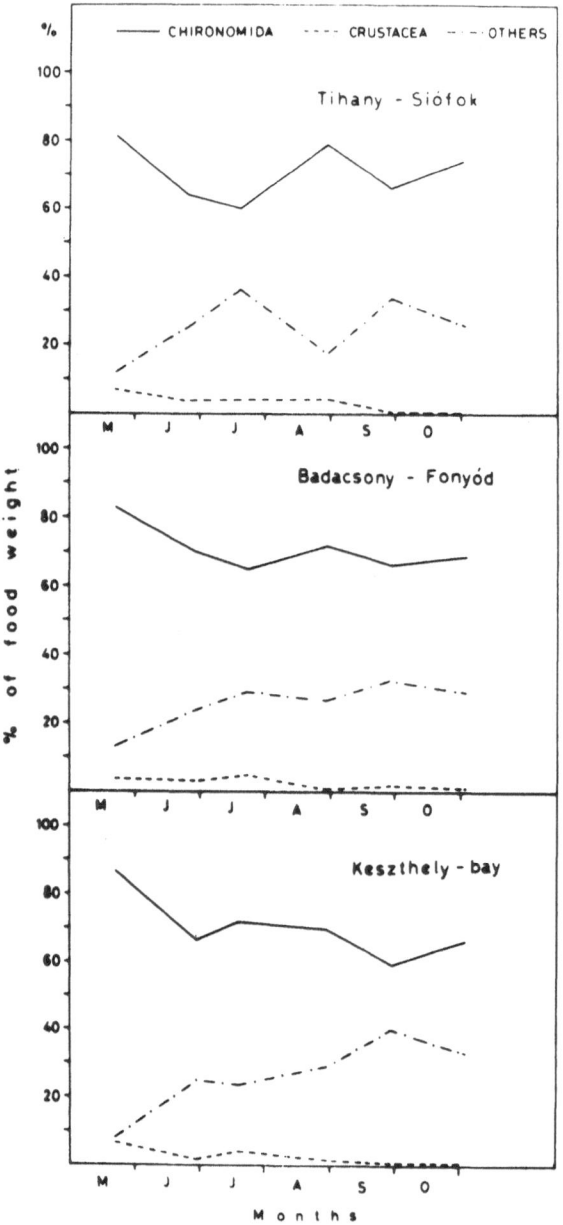

Fig. 4. Changes of the participation (in % weight) of the dominant animal food components of bream in 1977.

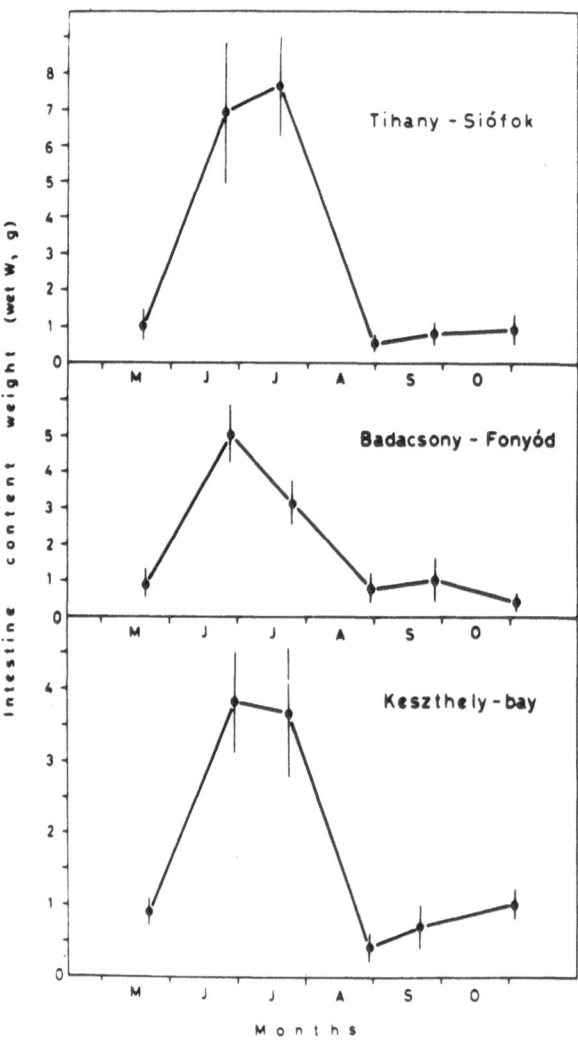

Fig. 5. Food intake of bream in different basins of Lake Balaton.

84

weight of food during the period examined. The bream consumed cladocerans (*Daphnia cucullata, D. hyalina, Diaphanosoma brachyurum, Alona affinis, Chydorus sphaericus*) and copepods (*Acanthocyclops vernalis, Mesocyclops leuckarti, Cyclops vicinus*) almost to the same extent. The frequency of occurrence in the alimentary tracts of bream varied from 20 to 100%.

These characteristics of the feeding habits of the mature bream are similar to findings of several authors (Kogan, 1963; Prejs, 1973; Melnichuk, 1975) who demonstrated that chironomid larvae were the dominant food component of bream, usually between 50 to 70% by weight. This is

evident because many authors agree with the general view that as bream grow older, they move to deeper feeding grounds and feed on bigger organisms (Backiel & Zawisza, 1968).

According to Karzinkin (1952) old bream find their food even under a 15 cm layer of mud. In Lake Balaton chironomid larvae move down even to 20 cm depth in the mud (Dévai, pers. comm.).

The dynamics of food intake of bream were similar in the three basins of Lake Balaton (Fig. 5). During the spawning period, May–June, the rate of feeding was small, the intestines containing about 1 g of food. The food intake reached a maximum in the summer months, July–August

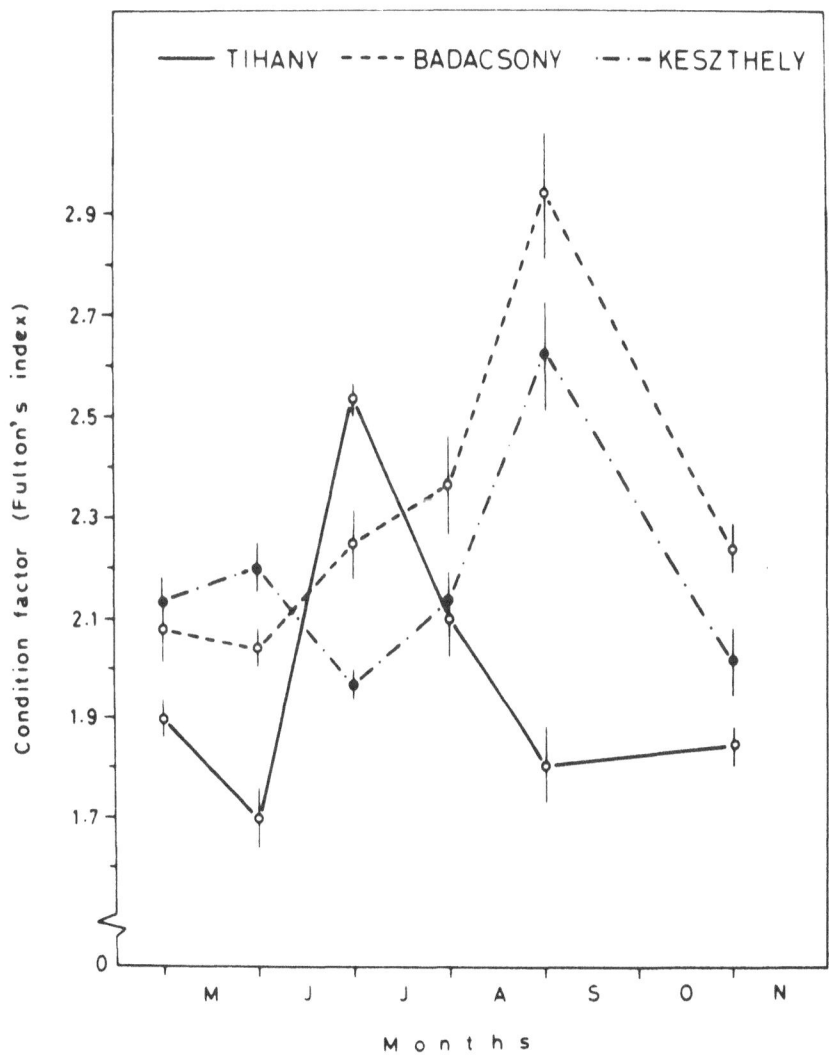

Fig. 6. Changes in the condition of bream living in different areas of Lake Balaton. Mean values and the range of Fulton's index in 1977.

(average 4 to 8 g/intestine) and there was a remarkable decrease in autumn when the food intake did not exceed the values observed in spring.

The condition of bream living in the Badacsony-Fonyód area and in the Keszthely-bay increased slightly from May to the end of August. From the beginning of August there was a sudden increase in the condition of bream which decreased again to a spring level (Fig. 6). The condition at the Tihany-Siófok area had another tendency: decreasing in June, increasing suddenly in July and increasing again in autumn.

The above changes in the condition of bream might be caused primarily by the food supply. The good condition of bream can be partly explained by the increased exploitation by fishing heavily influencing the population dynamics and partly by the fast eutrophication of Lake Balaton (Bíró & Garádi, 1974).

The rates of food consumption of bream and data of Bíró & Garádi (1974) on the abundance of adult non-piscivorous fish in Lake Balaton allowed us to do some calculations. According to these the bream consumed during the period examined about 30 to 35% of the standing crop of chironomids. The zooplankton was less affected as the adult bream consumed only 4 to 6% of the standing crop of zooplankton.

References

Backiel, T. & Zawisza, J. 1968. Synopsis of biological data of bream, Abramis brama (Linnaeus, 1958). FAO Fisheries Synopsis No. 36: 3:29–3:31.

Bíró, P. & Garádi, P. 1974. Investigations on the growth and population-structure of bream (Abramis brama L.) at different areas of Lake Balaton. The assessment of mortality and production. Annal. Biol. Tihany 41: 153–179.

Entz, B. 1954. Production biological problems of Lake Balaton. MTA Biol. és Orvosi Tud. Oszt. Közl. 5: 433–461. (Hung.).

Entz, B. 1965. Untersuchungen an Larven von Chironomus plumosus MEIG. im Benthos des Balatonsees in den Jahren 1964–1965. Annal. Biol. Tihany 32: 129–139.

Entz, B. & Lukacsovics, F. 1957. Untersuchungen im Winterhalbjahr und einigen Balatonsee-Fischen zwecks feststellung ihrer Ernährung-, Wachstums- und Vermehrungsumstände. Annal. Biol. Tihany 24: 71–86. (Hung., Germ. Summ.).

Herodek, S. 1977. Recent results of phytoplankton research in Lake Balaton. Annal. Biol. Tihany 44: 181–198.

Hillbricht-Ilkowska, A., Prejs, A. & Weglenska, T. 1973. Experimentally increased fish stock in the pond type Lake Warniak. VIII. Approximate assessment of the utilization by fish of the biomass and production of zooplankton. Ekol. Pol. 34: 553–562.

Hunt, B. P. 1960. Digestion rate and food consumption of Florida gar, warmouth and largemouth bass. Trans. Amer. Fish. Soc. 89: 206–211.

Kajak, Z. 1960. Dynamika liczebnosci Tendipedidae bentosowych na terenie mulistych odcinków lachy Konfederata. Ekol. Pol. A, 8: 229–260. (Pol., Eng. summ.).

Kajak, Z. & Dusoge, K. 1973. Experimentally increased fish stock in the pond type Lake Warniak. IX. Numbers and biomass of bottom fauna. Ekol. Pol. 35, 563–573.

Karzinkin, G. S. 1952. Osnovy biologicheskoi produktivnosti vodoemov. Moskva, Pischepromizdat, 402 p. (Ru.).

Kogan, A. V. 1963. On diurnal diet and rhythm of feeding in Abramis brama, L., of the Tsimlanskoe Reservoir. Vopr. Ikhtiol. 3: 319–325. (Ru., Eng. summ.).

Melnichuk, G. L. 1975. Ecology of feeding, food requirements and energy balance of the Dnieper Reservoir young fishes. Leningrad, Izv. GosNIORKH, 288 p.

Nauwerck, A. 1963. Die Beziehungen zwischen Zooplankton und Phytoplankton im See Erken. Symb. Bot. Upsal. 17: 1–163.

Oláh, J. 1976. Energy transformation by Tanypus punctipennis Meig. (Chironomidae) in Lake Balaton. Annal. Biol. Tihany 43: 83–92.

Ponyi, J. E. 1975. The biomass of zooplankton in Lake Balaton. Symp. Biol. Hung. 15: 215–224.

Prejs, A. 1973. Experimentally increased fish stock in the pond type Lake Warniak. IV. Feeding of introduced and autochtonous non-predatory fish. Ekol. Pol. 30: 465–505.

Sebestyén, O. 1958. Quantitative plankton studies on Lake Balaton. IX. Summary of the biomass studies. Annal. Biol. Tihany 25: 281–292.

Tátrai, I. 1977. Feeding habits and energy utilization by ruffe (Acerina cernua L.) in Lake Balaton. Ph.D. thesis, Tihany, 79 p. (Hung.).

Wojcik, S. 1970. Próba oceny ilosci larw Chironomidae w jeziorze Warniak. Acta Hydrobiol. 12: 309–322. (Pol., Eng. summ.).

Wojcik-Migala, I. 1965. Wplyw narybku karpi na dynamike fauny dennej. Roczn. Nauk. roln. 86: 215–227. (Pol., Eng. summ.).

4. MISCELLANEOUS

4.1 Chemical Aspects

PHYSICO-CHEMICAL ASPECTS OF ALKALINE PONDS IN YUGOSLAVIA

G. PETROVIĆ

Abstract

Physico-chemical aspects of shallow alkaline ponds (about 75) were studied for four years in the Pannonian plain in Yugoslavia. The paper describes results on the seasonal variations of their salinity, major ion composition, their chemical types, expressed as the percentage of equivalent sum of total cations or anions, their geographical distribution, as well as their similarity to the ionic composition of lakes and ponds in the Hungarian and Austrian part of the Pannonian plain.

Introduction

Alkaline salt lakes are dynamic systems. They respond rapidly to changes in external conditions, mainly climatic and tectonic. The chemical and mineralogical aspects of salt lakes are inextricably linked through the hydrological processes. Saline lake basins are not simple evaporating dishes but complex hydrological systems with spring, stream and groundwater inflow that is related to patterns of sediment packing and to the climatic and tectonic régimes. Saline lakes help us to understand processes which are important in the evolution of all natural waters, including those exposed to less extreme conditions (Eugster & Hardie, 1978).

Unfortunately, with some exceptions, they are characteristic of arid regions and generally accessible only with difficulty for continuous study (Williams, 1978).

Study area

Shallow astatic water bodies occur in north-east Yugoslavia, in a part of the Pannonian plain called Vojvodina covering an area of some 2.200,000 ha. They are usually in relief depressions and valleys of the Danube and Tisa and their tributaries. The lake areas are different in size ranging from less than 1 ha to 400 ha, their depth being usually below 50 cm, although some, like lake Palić, may have a depth of 8 m. The shallow lakes are temporary, all others are more or less permanent. In summer they undergo intensive evaporation and often dry out completely. During the winter they are frozen.

The present climate is temperate with dry summers and severe cold winters. Annual rainfall ranges between 550 and 650 mm, average monthly temperature varies between $-2°C$ and $22°C$ while the average annual temperature is $11°C$.

Tchernozeme is the dominant soil type. Alkaline soils also occur to some extent in the vicinity of the studied waters. Solonetz is the most widespread type of soil while solonchaks and soloti are seldom found. With regard to the salt composition in the solonchaks of Vojvodina, they are mainly sodium-chloride. Sulphate solonchaks are rare. The salt content in solonchaks usually ranges from 3–10‰, maximum 20‰. The saline soils have been formed by accumulation of salts from the underground waters.

The geological evolution of the Pannonian plain is that of a desiccated basin in which surface and underground waters have played a major part in the shaping and dynamics.

Physico-chemical aspects

Physico-chemical aspects of shallow alkaline lakes and ponds (about 75) were studied for four years with later sporadic visits (1950–1954, 1964). In the examined water bodies we assessed: salinity and its seasonal variations, major ion composition, basic chemical types expressed as the percentage of equivalent sum of total cations and anions, their geographical distribution, as well as their similarity to the ionic composition of lakes and ponds in other arid regions of the world.

Colour of the examined waters varies from green to different shades of green and brown, depending on the humus and other coloured organic matter deriving from decomposing reeds and other plant detritus. The other group of water bodies are silverish grey which is due to inorganic turbidity caused by very fine clay particles. The high turbidity is largely a function of wind activity, **shallow depth and nature of the sediments in the** basins. The rate of deposition is very small, particularly with colloidal suspensions. Thus, in one lake, 80 cm³ of sediments per litre were deposited after 7 days which gives evidence of the minimum light penetration through the surface layers of the water mass.

Salinity varies from 187–13.340 mg l⁻¹. Due to the small volume of these water bodies the fluctuations of salinity are very large both annually and over the course of many years. Total dissolved solids are the highest in summer and winter. During the wetter months in the majority of the waters the salinity does not exceed 1000 mg l⁻¹ but it increases a great deal during the drier months, particularly in small shallow water bodies, reaching amounts of between 2000 and 5000 (Table 1). Total solids exceeding 5000 were found in five lakes but only exceeding 10.000 in two lakes (Table 1). The source of the salts are (a) salty underground waters, (b) surrounding saline soils in the vicinity of the examined water bodies, (c) evaporation and freezing in winter.

All the lakes are alkaline, their pH seldom

Table 1. Seasonal variation of salinity in mg l⁻¹.

Lake	I	III	VII
		1952	
Opovo		344	656
Jezero I		886	1632
Jezero II		988	2872
Krvavo		1938	4735
Ridjice		2680	4308
Medura		2461	5364
Slatina	9006	1083	7712
Pečena Slatina	12731	2723	10240
Rusanda		4338	13340

being below 8, usually between 8.5 and 9.5, and sometimes over 10. The highest pH value was recorded in the waters with a rich content of carbonates. The carbonate content varies from 2.2 to 191.0 mg l⁻¹ CO_3. A considerable number of these water bodies are outstanding for their remarkable hardness (over 30°) due to magnesium and to some extent to calcium salts.

The Mg: Ca ratio is rarely below 1 and sometimes reaching 15. The waters rich in sodium salts have an insignificant hardness. Total hardness in the examined waters varies from 1.3–53°. The maximum concentration of magnesium was 220 mg l⁻¹ Mg and the maximum concentration of calcium was 136 mg l⁻¹ Ca. The lower calcium content may be attributed to the high pH values followed by the deposition of $CaCO_3$ and to the evaporation followed by deposition of $CaCO_3$ and $CaSO_4 \cdot 2H_2O$ (Hutchinson, 1957). Analyses of water extracts of sediments was carried out in some of the lakes and these always showed a higher calcium and sulphate content in comparison with the water (Petrović, 1956).

Total alkalinity varies according to the water body from 1.61–62.2 mval per litre. With increasing pH values the carbonate alkalinity increases on account of the bicarbonate alkalinity. The chloride content also varies within wide limits between different water bodies (4–2688 mg l⁻¹ Cl). Maximum values are found in the sodium and sulphate rich waters but not to the same extent in the bicarbonate rich waters. The chloride coefficient (total solids chloride) varies in the basins

Table 2. Minimum, maximum amd mean values of the most important chemical parameters in the examined ponds.

	min.	max.	mean.
salinity mg l^{-1}	187	13340	2041
alkalinity mval l^{-1}	1,61	62,20	9,44
carbonate hardness°	4,50	174,16	26,46
carbonate mg $l^{-1} CO_3$	2,2	191	38
chloride mg l^{-1} Cl	4	2688	367
sulphate mg $l^{-1} SO_4$	9	2024	256
calcium mg l^{-1} Ca	2,8	136,8	25,6
magnesium mg l^{-1} Mg	2,9	186	49
potassium permanganate consumption mg $l^{-1} KMnO_4$	20,4	416	55,6

Table 3. Major ion concentration in some of the examined ponds. Values A = mg l^{-1}; B = percentages of equivalent sum of total cations or anions.

Pond [+]		Salinity	Ca	Mg	Na+K	HCO_3+CO_3	SO_4	Cl
1	A	896	9	21	283	599	99	91
	B	-	3	12	85	68	14	18
2	A	865	33	50	127	549	56	40
	B	-	19	36	45	80	10	10
3	A	1065	13	15	261	675	42	46
	B	-	5	10	85	83	10	7
4	A	895	18	60	129	625	30	22
	B	-	8	42	50	89	6	5
5	A	622	26	48	150	437	99	91
	B	-	11	34	55	61	17	22
6	A	988	10	52	399	844	13	40
	B	-	3	27	70	91	2	7
7	A	746	64	288	481	634	227	93
	B	-	10	33	57	67	23	10
8	A	1314	42	149	165	612	386	135
	B	-	10	56	34	48	36	16
9	A	1176	100	91	135	424	400	75
	B	-	28	41	31	39	49	12
10	A	7451	16	11	279	2304	2024	1480
	B	-	1	1	98	31	34	35
11	A	638	24	22	193	187	31	274
	B	-	11	16	73	28	7	65

[+] Names of the ponds designated by the figures are found in table 7.

examined from 3–42. The sulphate content varies from 9–2205 mg $l^{-1} SO_4$. The sulphate coefficient (total solids/sulphate) ranges from 3–75.

Many of these lakes are heavily polluted as they are situated in inhabited areas, in settlements. The $KMnO_4$–consumption varies from 20–416 mg l^{-1} $KMnO_4$. Concentration of inorganic phosphorus ranged from 100–2000 μg/l P. The maximum values were in very alkaline waters with a high salinity (Table 2).

Chemical types of lakes

Since the salinity of these shallow lakes is very variable depending on the seasons and years the major ion composition (Tables 3–6) has been determined for all the lakes and chemical types according to the percentage of equivalent sum of total cations or anions exceeding 25% (Küfferath, 1952) (Table 7). It is important to emphasize that ionic proportions remain remarkably constant within a wide range of salt concentration but the pattern of ionic predominance is however very different even in closely situated lakes.

If the alkaline lakes of closed basins are best classified by their anion content, then the extreme water types in Yugoslavia are carbonate (or bicarbonate), sulphate and chloride waters. All possible intermediates exist. Among these alkaline lakes 23 chemical types of water have been recognized with a predominance of the Na—HCO_3 and Na—Mg—HCO_3 types, whereas the Ca—HCO_3 type

occurs only seldom and is the exception (Fig. 1). Such a diversity of chemical types of waters in Vojvodina is due to the geology of the region, its arid climate, the diversity of chemical types in the surrounding soils, the high level of the underground salt water and the shallow beds of two major rivers – the Danube and the Tisa which frequently flood the region. As is evident from Fig. 1, calcium occurs as a dominant ion only in four, magnesium in six and sodium in thirteen types. Sodium is a dominant cation in 52 lakes, its content reaching 93–98% of the cation sum in nine of the lakes. The highest sodium concentration was established in the following types: Na—Cl—SO_4—CO_3, Na—Cl—HCO_3 and Na—Cl.

Generally the bicarbonate-carbonate complex makes up a major portion of anions in many of

Table 4. Major ion concentration in some of the examined ponds. Values A = mg l⁻¹; B = percentages of equivalent sum of total cations or anions.

Pond[+]		Salinity	Ca	Mg	Na+K	HCO$_3$+CO$_3$	SO$_4$	Cl
12	A	2723	9	41	940	643	314	1038
	B	-	1	7	92	23	14	63
13	A	570	30	48	124	550	19	53
	B	-	14	36	50	82	4	14
14	A	994	49	47	254	550	212	140
	B	-	14	22	64	52	25	23
15	A	510	38	50	37	346	72	16
	B	-	25	54	21	74	20	6
16	A	1682	10	11	531	512	116	484
	B	-	2	4	94	34	11	55
17	A	4584	8	126	1608	2129	45	1592
	B	-	1	13	86	43	1	56
18	A	656	20	22	197	305	293	10
	B	-	9	11	80	44	54	2
19	A	2511	45	57	745	1024	600	360
	B	-	6	12	82	42	32	26
20	A	996	41	77	218	756	143	82
	B	-	12	36	52	70	17	13
21	A	606	38	66	107	677	14	24
	B	-	16	45	39	92	2	6
22	A	2571	17	14	930	750	670	576
	B	-	2	3	95	29	33	38

+) Names of the ponds designated by the figures are found in table 7.

Table 5. Major ion concentration in some of the examined ponds. Values A = mg l⁻¹; B = percentages of equivalent sum of total cations or anions.

Pond[+]		Salinity	Ca	Mg	Na+K	HCO$_3$+CO$_3$	SO$_4$	Cl
23	A	786	12	8	233	500	67	64
	B	-	5	6	89	72	12	16
24	A	300	28	12	59	199	48	40
	B	-	15	35	50	67	11	22
25	A	944	46	64	174	624	80	116
	B	-	15	35	50	67	11	22
26	A	1700	16	30	933	924	78	94
	B	-	4	13	83	78	8	14
27	A	1246	11	9	453	573	71	360
	B	-	3	4	93	45	7	48
28	A	187	23	9	31	177	13	4
	B		36	23	41	88	8	4
29	A	626	14	50	94	574	8	55
	B	-	20	37	43	85	1	14
30	A	426	34	24	55	268	37	32
	B	-	28	32	40	72	13	15
31	A	526	36	31	162	347	49	75
	B	-	20	30	50	64	12	24
32	A	940	11	4	285	665	59	74
	B	-	7	6	87	76	9	15
33	A	1414	4	10	453	781	222	119
	B	-	1	4	95	62	22	16

+) Names of the ponds designated by the figures are found in table 7.

the lakes and it is of major importance. Only three sodium chloride lakes are present.

Geographical distribution of chemical types

It is of interest to search for any regularity in the geographical distribution of chemical types. As is evident from the map (Fig. 1) the Na—HCO$_3$ is widespread in Vojvodina and it is characteristic of that region. In the region between the Sava and the Danube there are besides the Na—HCO$_3$ also the Mg—HCO$_3$ and Ca—HCO$_3$ types the latter occurring in the flood area of the river Sava. In the area between the Danube and Tisa rivers, in which the largest number of lakes occur, the most frequent types are: Na—Mg—HCO$_3$, Na HCO$_3$ and Mg—Na—HCO$_3$. Sulphate as a dominant

anion occurs in the northern part near Subotica and Ruski Krstur. There is a frequent combination of the Na—Mg and Mg—Na cations. In the area east of the Tisa the climate is more arid. There is a minor number of lakes but they exhibit the highest salinity, and the highest concentrations of sodium, chloride and sulphate. The Na—Cl, Na—Cl—HCO$_3$ and Na—Cl—SO$_4$—HCO$_3$ occur only in that part of Vojvodina.

There is a considerable number of Artesian wells in Vojvodina. According to Živković (1952) in the underground waters of Vojvodina besides Na—HCO$_3$ there are also Na—SO$_4$, Na—NCO$_3$—Cl—SO$_4$, Na—Ca—HCO$_3$ and Ca—Na—Cl—SO$_4$ types. Chloride anions only occur in the underground waters of the east part of the Tisa. The following basic conclusions have been reached. In

92

Table 6. Major ion concentration in some of the examined ponds. Values $A = mg\ l^{-1}$; $B =$ percentages of equivalent sum of total cations or anions.

Ponds[+]		Salinity	Ca	Mg	Na+K	HCO_3+CO_3	SO_4	Cl
34	A	263	43	32	23	320	27	10
	B	–	35	41	24	86	9	5
35	A	2818	48	189	643	875	1015	372
	B	–	6	33	61	31	46	23
36	A	1059	43	43	296	1041	21	40
	B	–	12	19	69	92	2	6
37	A	751	44	78	95	426	151	94
	B	–	18	50	32	54	25	21
38	A	5364	14	42	1927	3795	504	899
	B	–	1	3	96	63	11	26
39	A	842	9	21	292	562	147	92
	B	–	3	12	85	62	21	17
40	A	384	76	20	37	349	43	14
	B	–	55	22	23	81	13	6
41	A	885	17	20	308	872	73	28
	B	–	5	10	85	86	9	5
42	A	865	33	50	127	580	56	40
	B	–	15	36	49	80	10	10
43	A	550	17	30	111	299	60	72
	B	–	11	30	59	58	27	15
44	A	597	44	17	202	341	23	34
	B	–	20	13	67	86	6	8
45	A	989	16	25	121	962	79	24
	B	–	7	30	55	87	9	4

[+] Names of the ponds designated by the figures are found in table 7.

Table 7. Names of the ponds designated by the figures in the tables 3, 4, 5 and 6 and their chemical types determined according Küfferath (1952).

1. Slano	$Na-HCO_3$	
2. Bara kod Medure	$Na-HCO_3$	
3. Bela bara	$Na-HCO_3$	
4. Banja S. Miletić	$Na-Mg-HCO_3$	
5. Bara-Sevkerin	$Na-Mg-HCO_3$	
6. Jezero I	$Na-Mg-HCO_3$	
7. Stanišić	$Na-Mg-HCO_3$	
8. Odžaci	$Mg-Na-HCO_3-SO_4$	
9. Doroslovo	$Mg-Na-Ca-SO_4-HCO_3$	
10. Rusanda	$Na-Cl-SO_4-HCO_3$	
11. Bara II S	$Na-Cl-HCO_3$	
12. P. Slatina	$Na-Cl$	
13. Vilovo III	$Na-Mg-HCO_3$	
14. S. Sivac	$Na-HCO_3-SO_4$	
15. Opovo I	$Mg-Ca-HCO_3$	
16. Slatina	$Na-Cl-HCO_3$	
17. Slatina B.	$Na-Cl-HCO_3$	
18. Bara-Krstur	$Na-SO_4-HCO_3$	
19. Krvavo	$Na-HCO_3-SO_4-Cl$	
20. Ludaš	$Na-Mg-HCO_3$	
21. Kelebija	$Mg-Na-HCO_3$	
22. Okanj	$Na-Cl-SO_4-HCO_3$	
23. Opovo II	$Na-HCO_3$	
24. Matejski brod	$Na-Ca-HCO_3$	
25. Čonoplja	$Na-Mg-HCO_3$
26. R. rit	$Na-HCO_3$
27. Rakitaš	$Na-Cl-HCO_3$
28. Kumane	$Na-Ca-HCO_3$
29. Filipovo	$Na-Mg-HCO_3$
30. Žabalj	$Na-Mg-HCO_3$
31. Lalići	$Na-Mg-HCO_3$
32. Sv. Ivan	$Na-HCO_3$
33. Bosut	$Mg-Ca-HCO_3$
34. Palić	$Na-Mg-SO_4-HCO_3$
35. Kruševljevo	$Na-HCO_3$
36. M. Tisa	$Mg-Na-HCO_3-SO_4$
37. Medura	$Na-HCO_3-Cl$
38. Kula	$Na-HCO_3$
39. Obedska bara	$Ca-HCO_3$
40. Jakovo	$Na-HCO_3$
41. Jezero II	$Na-Mg-HCO_3$
42. Bara II M	$Na-Mg-HCO_3$
43. Bara II K	$Na-Mg-HCO_3$
44. Bara II J	$Na-HCO_3$
45. Jakovo II	$Na-Mg-HCO_3$

the region between the Danube and Tisa rivers, which has a wetter climate due to a higher rainfall and lower evaporation rate, the studied lake waters have lower salinity, higher concentration of bicarbonates and carbonates and a lower concentration of chlorides and sulphates.

Similarity of the lakes in Vojvodina to lakes in other arid regions

It is interesting to compare the chemical types of waters in Vojvodina with the corresponding types in the Hungarian part of the Pannonian plain and in Austria.

According to the data of Dvihally & Pony (1957) in the surroundings of Kistelek (the area between the Danube and Tisa) there are the following types of lakes: the $Na-CO_3-HCO_3$, $Na-Mg-HCO_3-CO_3$ and $Na-Mg-CO_3$. Woynarovich (1941), who investigated the stagnant waters in the area between Szeged and Budapest which includes fishponds and Neusiedlersee, it was established that, besides the above mentioned, the following types are present as well: $Na-HCO_3-Cl$, $Na-Ca-CO_3-HCO_3$, $Ca-Na-HCO_3-SO_4$ and $Mg-HCO_3$.

Almost all these types occur in the Yugoslav part of the Pannonian plain – Vojvodina.

We also found a strong similarity of the chemical types of waters in the Yugoslavian and Hungarian parts of the Pannonian plain with the results of Maucha (1969) on Hungarian lakes.

Results of a study of lakes in the Seewinkel in the Austrian part of the Pannonian plain, between Neusiedlsee and the Hungarian frontier, 80 km

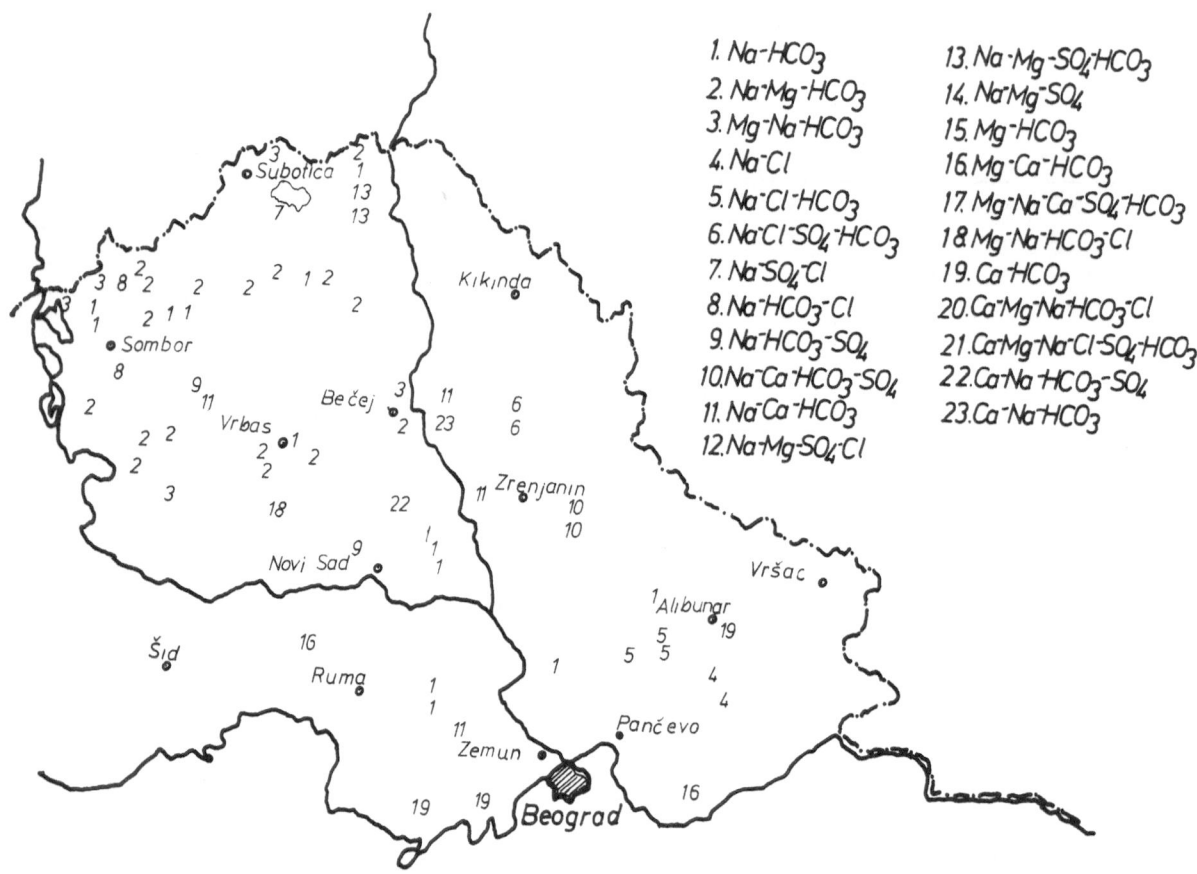

Fig. 1. A map of the investigated region. Geographical distribution of chemical types in Vojvodina.

Legend for map:

1. Na-HCO₃
2. Na-Mg-HCO₃
3. Mg-Na-HCO₃
4. Na-Cl
5. Na-Cl-HCO₃
6. Na-Cl-SO₄-HCO₃
7. Na-SO₄-Cl
8. Na-HCO₃-Cl
9. Na-HCO₃-SO₄
10. Na-Ca-HCO₃-SO₄
11. Na-Ca-HCO₃
12. Na-Mg-SO₄-Cl
13. Na-Mg-SO₄-HCO₃
14. Na-Mg-SO₄
15. Mg-HCO₃
16. Mg-Ca-HCO₃
17. Mg-Na-Ca-SO₄-HCO₃
18. Mg-Na-HCO₃-Cl
19. Ca-HCO₃
20. Ca-Mg-Na-HCO₃-Cl
21. Ca-Mg-Na-Cl-SO₄-HCO₃
22. Ca-Na-HCO₃-SO₄
23. Ca-Na-HCO₃

east of Vienna, Knie (1958) and Löffler (1959) reveal a similar ionic composition and concentration ranges of total dissolved solids.

According to the data of Löffler (1959) there are the following types of lakes: Na—HCO₃, Na—Mg—HCO₃, Mg—Na—HCO₃—SO₄, Na—Mg—HCO₃, Na—Mg—HCO₃—SO₄ (the corresponding data transformed into the equivalent percentages by Petrović). The electrolytic conductivity of the water bodies in the Seewinkel ranges from 940–14.930 μS.

If the alkaline waters in Austria (besides the Hungarian sodium bicarbonate type of lakes) reflect the essential state of the alkaline water types in Europe, with a summer alkalinity value of 143 mval. per litre as the peak value for the continent (Löffler, 1959) then our studies of the saline lakes in Yugoslavia complete this picture.

References

Dvihally, Z. & Ponyi, J. 1957. Characterisierung der Natrongewässer in der Umgebung von Kistelek auf Grund ihrer chemischen Zusammensetzung und ihrer Crustacea-Fauna. Acta Biol. Acad. sci. hungar. 7: 349–363.

Eugster, P. H. & Hardie, A. L. 1978. Saline lakes. Lakes-chemistry, geology, physics. Springer-Verlag, New York, 363 pp.

Gorbov, A. F. 1951. Ueber die Zonalität der chemischen Zusammensetzung der Gewässer in der Kulundinskaja Steppe. Ber. Akad. Wiss. UdSSR, 81.

Knie, K. 1961. Ueber den Chemismus der Wasser im Seewinkel der Salzlackensteppe Oesterreichs. Verh. Internat. Verein. Limnol. 14: 1142–1143.

Küfferath, H. J. 1951. Representation graphique et classification chimique rationelle. Inst. roy. des sci. natur. de Belg. 27: 1–8.

Löffler, H. 1959. Zur Limnologie, Entomostraken – u. Rotatorien Fauna des Seewinkelgebietes (Burgenland, Oesterreich). Sitzungsberichte Oesterreichische Akademie den Wissenschaften, 168: 315–362.

Hutchinson, G. E., Pickford, E. G. & Schurmann, J. 1932. A contribution to the Hydrobiology of pans and other inland waters of South-Africa. Arch. Hydrobiol. 24: 1–154.

Hutchinson, G. E. 1957. A Treatise on Limnology. Vol. 1. John Wiley, New York. 1015 pp.

Maucha, R. 1949. Einige Gedanken zur Frage des Nährstoffhaushalts der Gewässer. Hydrobiology, 1: 225–337.

Nejgebauer, V. 1952. Činioci stvarnaja zemljišta u Vojvodini. Zbornik Matice srpeske, 2, Novi Sad.

Petrović, G. 1956. Types chimiques des eaux superficielles et souterraines dans la région pannonienne. Rapports, 1–2, VIe Congres International de la Science du Sol, 111–115. Paris.

Williams, W. D. 1978. Limnology of Victorian salt lakes. Verh. Internat. Verein. Limnol. 20: 1165–1174.

Woynarovich, E. 1941. Untersuchungen über die chemischen Eigenschaften einiger ungarischen Gewässer. Magyar. Biol. Kut. Munk. 13: 302–315.

Živković, M. 1952. Ispitivanje kvaliteta bunarske vode za navodnjavanje na teritoriji Vojvodine – Zemljište i biljka 1 (3). Beograd.

PHOSPHORUS CONCENTRATION AND ALGAL SPECIES COMPOSITION IN TWO EUTROPHIC RESERVOIRS IN SOUTHERN ENGLAND

R. E. YOUNGMAN & M. R. FARLEY

Abstract

Two adjacent 9 m deep man-made reservoirs were studied for two years. Both were filled with eutrophic River Thames water from the same point. Phosphorus concentrations in the two reservoirs differed greatly owing to differences in the management of the water bodies. Algal species composition was analysed in relation to phosphorus concentration. The value of phosphorus removal from such waters is discussed.

Introduction

Prior to 1967 the city of Oxford, in central southern England, had obtained its water supply by direct abstraction from the River Thames but in that year the first of two new reservoirs was brought into use. The second was completed and filled 10 years later. The storage volume of these reservoirs ensures that the water needs of the city can be met in drought years when the discharge of the river is insufficient.

The reservoirs

Farmoor I (RI) and Farmoor II (RII) reservoirs can be thought of as artificial lakes, created by building ring-shaped barriers. They are situated on the flood plain of the River Thames some 6 km west of Oxford. The site consists of alluvial gravels underlain by a thick and continuous mantle of Oxford Clay. The whole site was cleared down to the clay, and the gravels and some clay were used for the construction of the reservoir embank-

ments. These stand 9 m above the original ground level and top water level is 67 m A.O.D. The inner surfaces of the embankments are protected from wave action and erosion by concrete slabbing. In plan the reservoirs are approximately D-shaped. The straight side of the D is shared as a common embankment. The physical characteristics of the reservoirs are given in Table 1.

The reservoirs have no natural inflows other than precipitation and are filled by pumping from the Thames which at the point of abstraction is a fairly slow flowing river (average daily discharge 15 to 20 $m^3 s^{-1}$) which has passed through largely agricultural land. The town of Swindon (population c. 90 000) which is 45 km upstream contributes the only major treated sewage effluent (c. 0.4 $m^3 s^{-1}$). The remaining small towns and villages above Farmoor contribute a further 0.2 $m^3 s^{-1}$. Normally the reservoirs are maintained at top water level by pumping at a rate equivalent to the demand for water. The reservoirs become drawn down when river water cannot be pumped in because of low river discharge or when it is considered necessary to allow a pollutant to go past the intake in order to protect the supply.

Two water treatment works take their water from the reservoirs though if it should be necessary both can be supplied directly from the river by means of a reservoir bypass main. Water can be abstracted from RI through a valve tower at the eastern end which has outlets at three levels. The valve tower in RII is also at the eastern end.

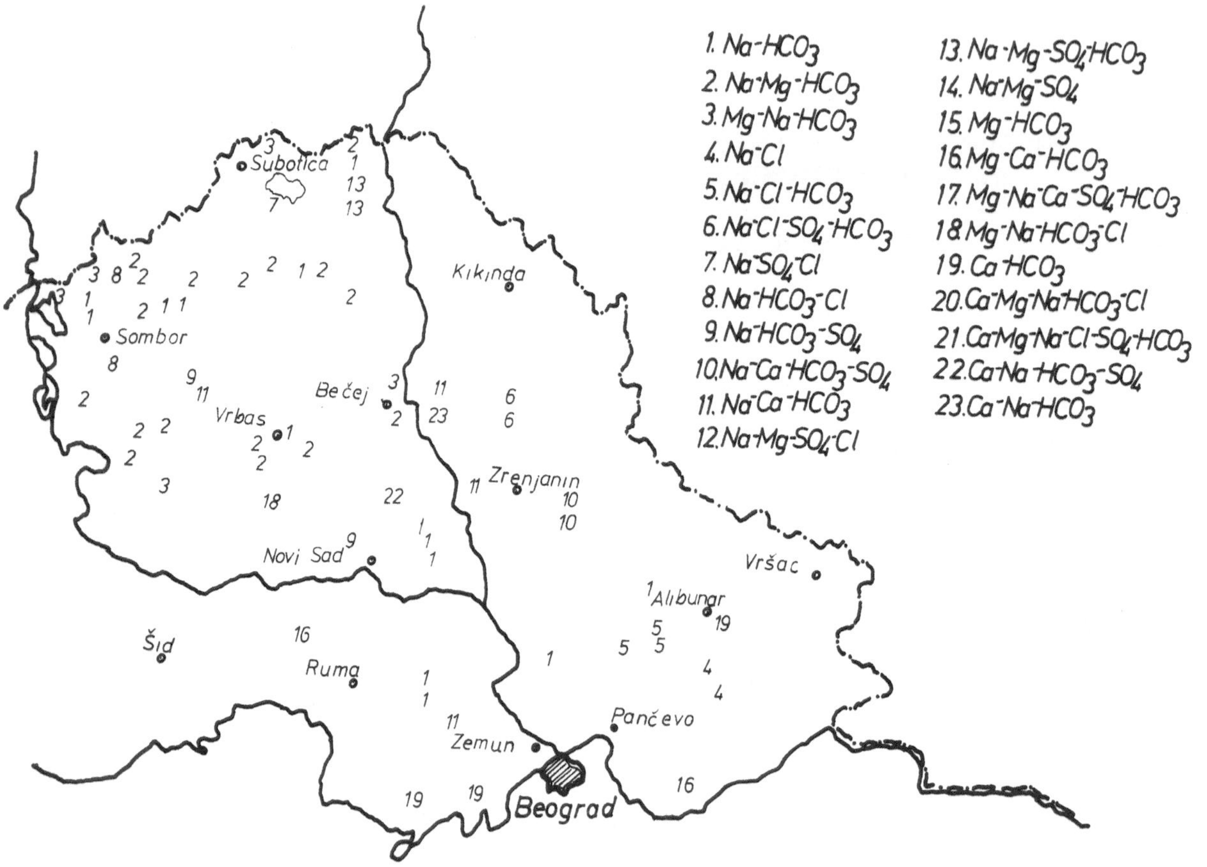
1. Na-HCO3
2. Na-Mg-HCO3
3. Mg-Na-HCO3
4. Na-Cl
5. Na-Cl-HCO3
6. Na-Cl-SO4-HCO3
7. Na-SO4-Cl
8. Na-HCO3-Cl
9. Na-HCO3-SO4
10. Na-Ca-HCO3-SO4
11. Na-Ca-HCO3
12. Na-Mg-SO4-Cl
13. Na-Mg-SO4-HCO3
14. Na-Mg-SO4
15. Mg-HCO3
16. Mg-Ca-HCO3
17. Mg-Na-Ca-SO4-HCO3
18. Mg-Na-HCO3-Cl
19. Ca-HCO3
20. Ca-Mg-Na-HCO3-Cl
21. Ca-Mg-Na-Cl-SO4-HCO3
22. Ca-Na-HCO3-SO4
23. Ca-Na-HCO3
Subotica
Kikinda
Sombor
Bečej
Vrbas
Zrenjanin
Novi Sad
Vršac
Alibunar
Šid
Ruma
Pančevo
Zemun
Beograd

Fig. 1. A map of the investigated region. Geographical distribution of chemical types in Vojvodina.

east of Vienna, Knie (1958) and Löffler (1959) reveal a similar ionic composition and concentration ranges of total dissolved solids.

According to the data of Löffler (1959) there are the following types of lakes: Na—HCO_3, Na—Mg—HCO_3, Mg—Na—HCO_3—SO_4, Na—Mg—HCO_3, Na—Mg—HCO_3—SO_4 (the corresponding data transformed into the equivalent percentages by Petrović). The electrolytic conductivity of the water bodies in the Seewinkel ranges from 940–14.930 μS.

If the alkaline waters in Austria (besides the Hungarian sodium bicarbonate type of lakes) reflect the essential state of the alkaline water types in Europe, with a summer alkalinity value of 143 mval. per litre as the peak value for the continent (Löffler, 1959) then our studies of the saline lakes in Yugoslavia complete this picture.

References

Dvihally, Z. & Ponyi, J. 1957. Characterisierung der Natrongewässer in der Umgebung von Kistelek auf Grund ihrer chemischen Zusammensetzung und ihrer Crustacea-Fauna. Acta Biol. Acad. sci. hungar. 7: 349–363.

Eugster, P. H. & Hardie, A. L. 1978. Saline lakes. Lakes-chemistry, geology, physics. Springer-Verlag, New York, 363 pp.

Gorbov, A. F. 1951. Ueber die Zonalität der chemischen Zusammensetzung der Gewässer in der Kulundinskaja Steppe. Ber. Akad. Wiss. UdSSR, 81.

Knie, K. 1961. Ueber den Chemismus der Wasser im Seewinkel der Salzlackensteppe Oesterreichs. Verh. Internat. Verein. Limnol. 14: 1142–1143.

Küfferath, H. J. 1951. Representation graphique et classification chimique rationelle. Inst. roy. des sci. natur. de Belg. 27: 1–8.

Löffler, H. 1959. Zur Limnologie, Entomostraken - u. Rotatorien Fauna des Seewinkelgebietes (Burgenland, Oesterreich). Sitzungsberichte Oesterreichische Akademie den Wissenschaften, 168: 315–362.

shaken sample through a Whatman GF/C glass fibre filter which had previously been well washed with at least 100 ml of distilled water. When phosphate was expected to be at levels below $0.010 \, \text{mg P} \cdot 1^{-1}$ the filter was washed with 1000 mls of distilled water.

All samples and sub-samples were stored in polyethylene bottles, those for chemical analysis at 4°C. The analytical methods used have been described elsewhere (Youngman, 1975).

Results

The Thames

The observed concentrations of nitrate nitrogen and dissolved orthophosphate phosphorus are shown in Figure 1. Clearly the river water is rich in these algal nutrients. Over the period the concentration of ortho-phosphate varied between 114 and $1560 \, \mu\text{g P} \cdot 1^{-1}$. Highest values occurred in autumn reflecting an earlier lack of rain to dilute the largely sewage-derived input. The autumn of 1978 was particularly dry but with heavy rain in early December dilution reduced concentrations to the usual winter levels.

Nitrate concentrations varied between 5.81 and $11.4 \, \text{mg N} \cdot 1^{-1}$ with the lowest values in late summer and the highest in winter. A marked increase in concentration followed the December rain suggesting that the extra nitrate entering the river at this time was in run-off of agricultural fertilizer.

Nutrient loadings

Table 2 shows the areal loadings of nitrate nitrogen and ortho-phosphate phosphorus to the reservoirs in 1978, and compares them with values for 1968 and with Vollenweider's (1968) threshold values for a 10 m deep eutrophic lake. Even in 1968 RI was highly eutrophic and over the decade the loadings have increased very considerably, largely as a result of the increased throughput of water. The reservoir must now rate as one of the most eutrophic water bodies in the world. Loadings to RII during 1978 were zero as no water was pumped into the reservoir.

The reservoirs

Figure 2 shows the concentrations in the two reservoirs of nitrate and ortho-phosphate phosphorus and demonstrates the large differences in chemical quality which resulted from the different operating regimes. Although bacterial denitrification and probably algal assimilation of nitrate took place in both reservoirs, concentrations in RII, which was denied an input of nitrate-rich river water, fell steadily until by August 1979 they had fallen below the limit of detection $(0.06 \, \text{mg N} \cdot 1^{-1})$ of the analytical method. As expected, concentrations in RI exhibited a seasonal pattern broadly similar to that of nitrate in the river. At the start of the period there was a difference between the reservoirs of only $0.01 \, \text{mg} \, 1^{-1}$, at the time of the maximum observed concentration in RI there was over six times more nitrate nitrogen in RI than RII and by August 1979 over eleven times.

Phosphorus concentrations in RII had by 24th July 1978 fallen below the limit of detection $(6 \, \mu\text{g P} \cdot 1^{-1})$ of the analytical method in use at that time. This was probably largely due to the assimilation of the phosphorus by the algae in the reservoir. Later analyses by a more sensitive method demonstrated that the concentration was frequently between 2 and $4 \, \mu\text{g P} \cdot 1^{-1}$ and on 2nd April 1979 that it fell to less than $0.5 \, \mu\text{g P} \cdot 1^{-1}$. However, during the winter of 1978/79 and in September 1979 reservoir phosphorus concentrations increased, reaching $49 \, \mu\text{g P} \cdot 1^{-1}$ on 22nd January 1979. The reasons for these increases are, at present, not known.

At the time of the phosphorus maximum in RI there was thirty-five times more than in RII. Algal growth in RI was very unlikely to have been limited by a phosphorus shortage whereas that in RII may well have been.

Both reservoirs exhibit a typical seasonal succession of algal chlorophyll (total pigment) maxima (Fig. 3). The large between-reservoir differences in the concentrations of nitrogen and phosphorus are clearly not reflected by the chlorophyll figures, indeed the maximum values for chlorophyll occurred in RII rather than RI. However, except in the first weeks of June and August 1979, the maxima in the reservoirs are not coincident. From July 1978 to July 1979 values in RII did not exceed $35 \, \mu\text{g} \, 1^{-1}$ though this level was frequently exceeded in RI. In the following section the organisms responsible for the chlorophyll

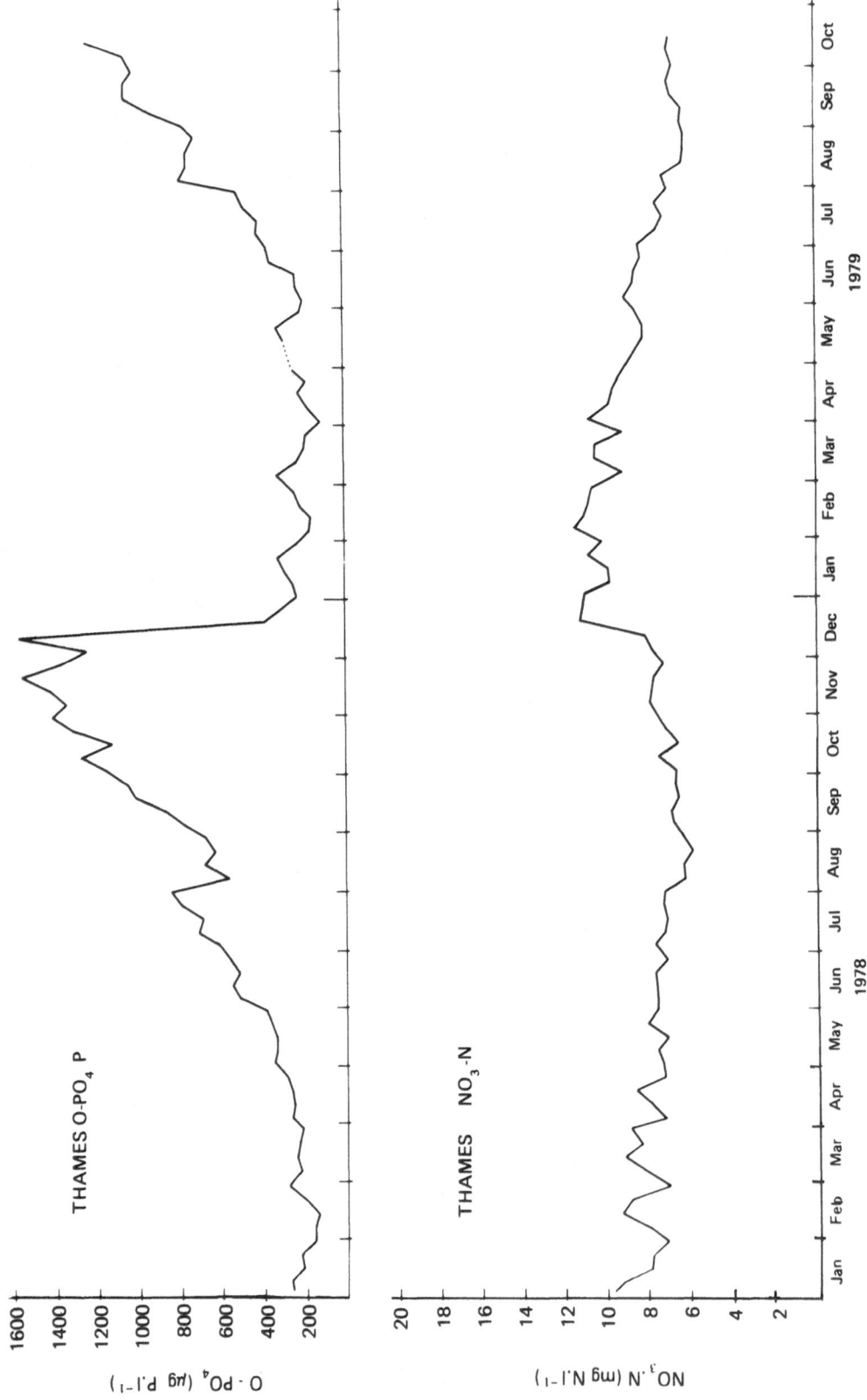

Fig. 1. Phosphorus and nitrogen concentrations in the Thames during 1978 and 1979.

100

Table 2. Areal loadings of phosphorus and nitrogen to Farmoor I and II reservoirs. Vollenweider's (1968) threshold values, for a 10 m deep eutrophic lake in parentheses

	Areal loading P(0.2)	g.m.$^{-2}$a^{-1} N(3.0)	Retention time days
Farmoor I (1968)	5	112	173
(1978)	37	425	59
Farmoor II (1978)	0	0	Infinite

maxima are investigated to determine whether there were any marked differences in the algal taxa occurring in the two reservoirs.

The algae

Diatoms

During the spring of 1978, when nitrate and phosphorus concentrations in the two reservoirs were broadly similar, the dominant diatom species were the same (Table 3) though *Stephanodiscus hantzschii* was present in only small numbers in RI. In 1979, when nutrient concentrations were very different, *S. hantzschii* was numerically the most abundant organism in RI and the population of *Asterionella* was small. Centric diatoms (*S. astraea* and *S. hantzschii*) were very poorly represented in RII where the dominants were *Asterionella* and *Fragilaria*. The latter had also been present in the autumn of 1978 (maximum count 4852 cells ml^{-1} on 4th September).

Silica concentration appeared to limit the growth and eventual maximum cell number of all the diatoms. In the winter of 1978–79 the maximum concentrations observed in RI and RII were 6.18 and 2.56 mg l^{-1} respectively. Pumping from the river and, probably, internal recycling from the sediments replenished the silica in RI but in RII only recycling could have occurred. This difference in concentration clearly influenced the maximum observed numbers of diatoms in the two reservoirs.

Green algae

Both reservoirs had populations of colonial green algae (Table 3) in the late spring and early summer of 1978. Similar crops were reported for RI

during 1968–1973 (Youngman, 1975). Unusually, these algae were unimportant in 1979. Only the unicellular green algae *Carteria*, in RI, and *Chlorella*, in RII, showed significant growth. As growth of colonial green algae did not occur in either reservoir their absence in this year cannot be attributed to the nitrogen or phosphorus concentrations. Other factors, including grazing by herbivorous zooplankters, may have been responsible. However, the growth of *Chlorella* in RII was probably limited by phosphorus since on 2nd April 1979 its concentration fell to less than 0.5 μg l^{-1}.

Dinoflagellates

Although present, no numerically abundant populations of dinoflagellates were observed in RI. In RII however both *Peridinium*, in 1978, and *Ceratium*, in 1979, occurred (Table 3). The outburst of the latter organism was particularly striking in that growth, which commenced in early June, lasted for more than four months, the maximum occurring in August.

Blue-green algae

Three species showed appreciable growth (Table 3). In 1978 the colonial organism *Microcystis* grew well in both reservoirs. Colonies were ultrasonicated so that standing crop could be expressed in terms of cell numbers. Unfortunately this procedure was not followed in 1979 and only counts of colonies were made (Table 4). From these, and the visual appearance of the water body, it is clear that in RI the *Microcystis* population was as abundant as in 1978. In fact this organism has typified the autumn plankton of this reservoir for the best part of a decade. An autumn bloom of blue-green algae did occur in RII in 1979 but the dominant organism was *Aphanizomenon*. This also occurred in RI in both years. *Aphanizomenon*, in contrast to *Microcystis*, is a nitrogen fixer and it occurred when nitrate concentrations in RII were very low. However, in this and previous years in RI it has often grown when nitrate was plentiful.

Discussion

The reduction of nutrient concentrations in RII, resulting from the cessation of pumping, has been

101

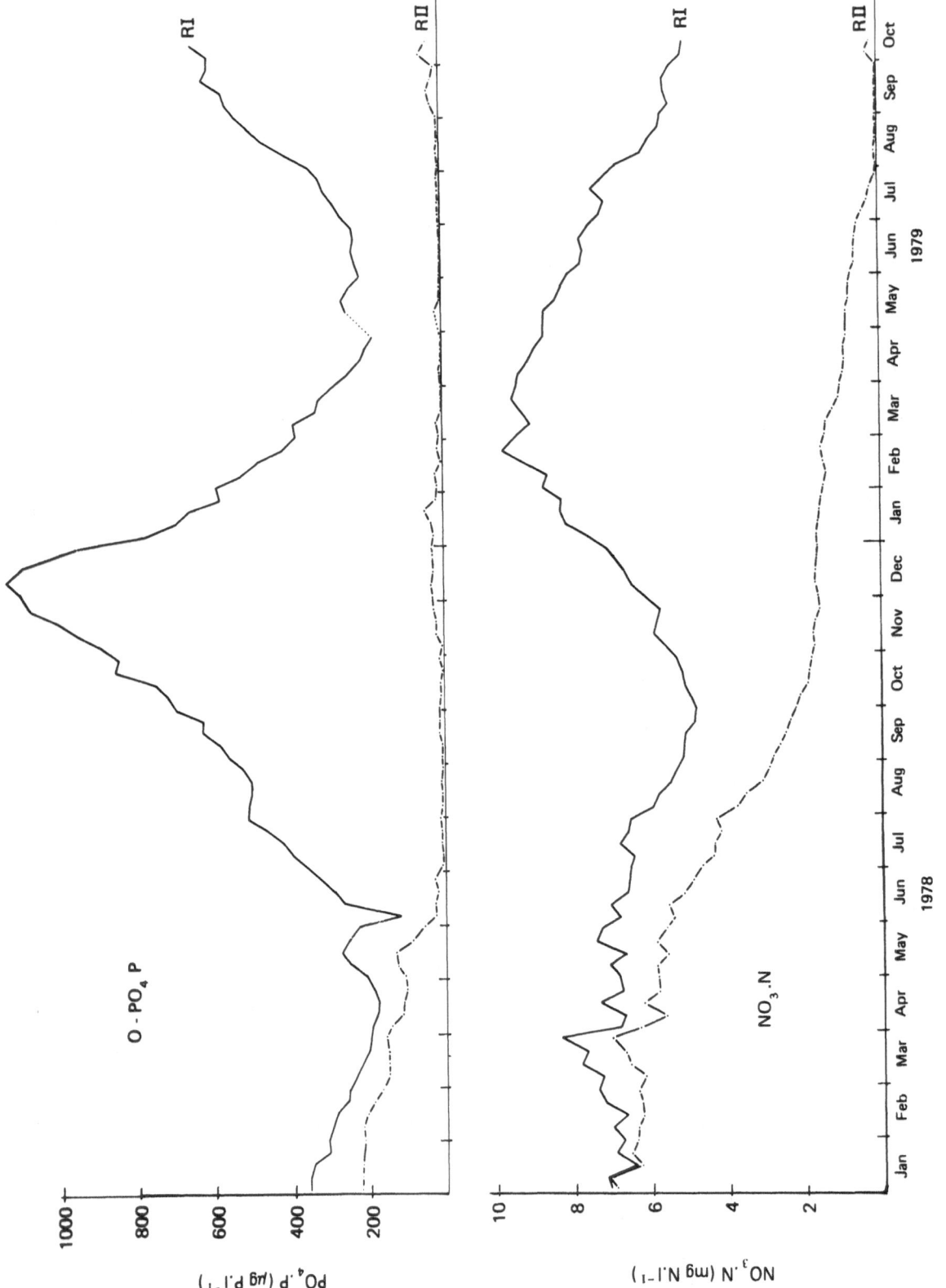

Fig. 2. Phosphorus and nitrogen concentrations in Farmoor I and II reservoirs during 1978 and 1979.

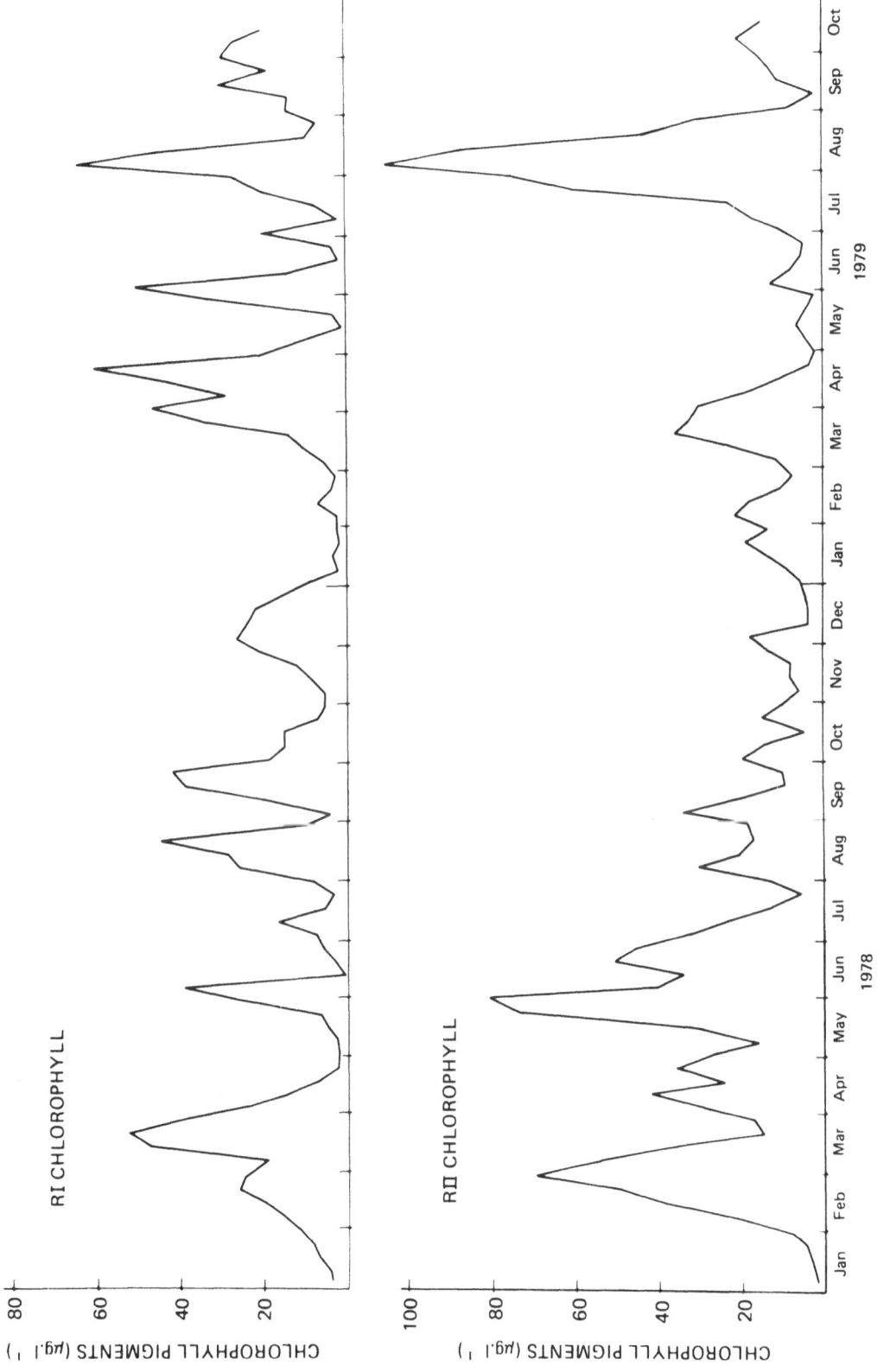

Fig. 3. Chlorophyll concentrations in Farmoor I and II reservoirs during 1978 and 1979.

103

Table 3. Maximum algal counts (Dates).

		1978		1979	
	Diatoms	RI	RII	RI	RII
*	Asterionella formosa Hass.	6732 (28/3)	3942 (13/2)	262 (17/4)	3252 (5/2)
*	Stephanodiscus astraea (Ehr) Grün.	584 (20/3)	385 (27/2)	484 (4/6)	2 (15/1)
*	Stephanodiscus hantzschii Grün.	1208 (13/3)	48219 (10/4)	81313 (23/4)	192 (26/3)
*	Fragilaria crotonensis Kitt.				4157 (22/1)
	Green algae				
o	Pandorina morum Bory.	1243 (5/6)			
o	Coelastrum microporum Näg.		557 (4/7)		
o	Dictyosphaerium pulchellum Wood		13310 (24/4)		
o	Oocystis sp.		2650 (30/5)		
*	Chlorella sp.			5395 (9/4)	210000 (19/3)
*	Carteria sp.			12408 (6/8)	
	Flagellates				
*	Peridinium sp.		232 (11/9)		
*	Ceratium hirundinella (O.F.M.) Schrank				501 (13/8)
	Blue-green algae				
-	Anabaena flos-aquae Bréb. ex Born et Flah	448 (14/8)	101 (2/5)	148 (25/6)	7 (2/7)
-	Aphanizomenon flos-aquae (L.) Ralfs	1216 (10/7)	80 (20/11)	2570 (2/7)	3212 (8/10)
*	Microcystis aeruginosa Kütz.	114229 (25/9)	88749 (23/10)	(See Table 4)	

* cells ml^{-1}
o colonies ml^{-1}
- filaments ml^{-1}

accompanied by significant changes in the algal flora. Centric diatoms have become less common and may have been replaced by *Fragilaria*, though *Asterionella* persists. Large crops of *Ceratium* have become a feature and very much reduced populations of *Microcystis* now occur.

These changes illustrate what might be achieved if phosphorus removal were applied to the Thames water before it was pumped to a reservoir not operated as a standing reserve. From operating experience at Farmoor (Crowe, 1974; Youngman, 1975) it can be seen that not all the changes are beneficial. The coagulation, sedimentation, rapid filtration (CSF) process used at Farmoor is especially suited to the removal of many small algae rather than fewer large algae (Collingwood, 1977). The more frequent occurrence of larger diatoms and *Ceratium* could be a problem rather than a benefit. But any reduction in the amount of *Microcystis* present in the stored water would clearly be of great benefit; water treatment at Farmoor has often been seriously interfered with when this organism has been present, largely because CSF becomes less efficient at pH values in the range 8.5 to 10.0 units which may be experienced during blooms of blue-green algae. During 1976, which was an unusual year because of a drought which existed over much of Europe, a bloom of *Microcystis* in RI persisted from May to October at an average chlorophyll concentration of 56 μg l^{-1}. Treatment difficulties brought about by the secretion of extracellular products by the alga continued for several months after the organism had disappeared from the water column. Similar, though less prolonged, difficulties have been experienced in other years.

The merits of phosphorus removal should be determined not only on the basis of reductions in nutrient concentration but also in the light of changes in amount and type of algal material. In

Table 4. Numbers per millilitre of *Microcystis* in 1979.

Date	Cells	RI Colonies small*	medium	large	Cells	RII Colonies small	medium	large
16.7	62	6	0.14	0.18	0	0	0	0
23.7	415	88	28	0.66	0	0	0	0
30.7	996	40	7	0.1	0	0	0	0
6.8	259	657	253	1.1	0	0	0	0
13.8	2552	56	22	0	0	0	0	0.2
20.8	3922	131	19	0.65	0	0	0	0
28.8	2643	38	5	1.15	0	0	0	0
3.9	8300	66	3	1.3	2081	0	0	0
10.9	9098	104	15	0.9	16	3	0	0
17.9	37529	79	46	4.7	286	39	0	0
24.9	6619	51	25	4.5	0	0	0.1	0
1.10	1535	111	18	4.0	0	16	0.4	0.01
8.10	768	87	33	3.8	56	19	1	0
15.10	4032	94	18	4.7	58	2	0.7	0

* Colony size was arbitrarily defined as

small < 30 µ diameter, medium 30-150 µ, large > 150 µ.

turn these must be considered in relation to the treatment process.

Collingwood (personal communication) has calculated that to remove 96% of the phosphorus entering Farmoor I from the Thames would require a treatment plant with facilities for coagulant dosing, sedimentation and filtration similar to those provided by the water-works. As a rough approximation therefore phosphorus removal would double the treatment costs. Such an increase is difficult to justify for highly eutrophic waters and this may often be the factor of most importance when phosphorus removal is being considered.

Acknowledgements

The authors wish to thank the Divisional Manager, Mr. R. G. Littlewood, and the staff of the Vales Division, Thames Water Authority for permission to work at Farmoor and for their cooperation. They also acknowledge the assistance of their colleagues at the Water Research Centre.

References

Collingwood, R. W., 1977. A survey of eutrophication in Britain and its effects on water supplies. Water Research Centre Technical Report TR 40. Medmenham, The Centre.

Crowe, P. J., 1974. The Farmoor source works in operation. J. Instn. Wat. Engrs., 28: 101–124.

Lack, T. J. & Collingwood, R. W., 1975. The control of reservoir water quality by engineering methods. Proceedings of the Water Research Centre Symposium on "The effects of storage on water quality". Medmenham, The Centre.

Lund, J. W. G., Kipling, C. & Le Cren, E. D., 1958. The inverted microscope method of estimating algal numbers and the statistical basis of estimations by counting. Hydrobiologia, 11, 143–170.

Lund, J. W. G. & Talling, J. F., 1957. Botanical limnological methods with special reference to the algae. Bot. Rev., 23, 489–583.

Talling, J. F., 1969. In: A manual on methods for measuring primary production in aquatic environments. Ed. R. A. Vollenweider. Oxford. Blackwell Scientific Publications. (pp. 22–25).

Vollenweider, R. A., 1968. Scientific fundamentals of the eutrophication of lakes and flowing waters with particular reference to nitrogen and phosphorus as factors in eutrophication. Techn. Rept. DAS/CS 1/68.27. Organization for Economic Co-operation and Development. Paris.

Water Research Association, 1971. Algal monitoring of water supply reservoirs and rivers. Technical Memorandum TM 63. Medmenham, The Association.

Youngman, R. E., 1975. Observations on Farmoor, a eutrophic reservoir in the upper Thames valley, during 1965–1973. Proceedings of the Water Research Centre Symposium on "The effects of storage on water quality", Medmenham, The Centre.

Youngman, R. E., 1978. The measurement of chlorophyll. Water Research Centre Technical Report TR 82. Medmenham, The Centre.

Youngman, R. E., Johnson, D. & Belling, P., 1973. Some initial studies on vitamins in relation to the phytoplankton crops in a eutrophic reservoir. Wat. Treat. Exam., 22: 233–249.

Youngman, R. E., Farley, M. R. & Johnson, D., 1976. Factors influencing phytoplankton growth and succession in Farmoor Reservoir. Freshwat. Biol., 6: 253–263.

FARMOOR I AND II, SOUTHERN ENGLAND

BACKGROUND DATA

Latitude: 51°45'N

origin: man made reservoirs

	I	II
Longitude: 1°21'W		
Altitude: 67 m	z (m) 11.3	12.0
1 (km) not appropriate-	$\bar{z}$ (m) 9.0	9.0
b (km) man made reservoirs	$\bar{z}/z$ 0.8	0.75
L (km) Length of shore line:	V (km³) 0.0046	0.0093
I 2.5 km, II 3.9 km	A (km²) 0.506	1.02
L_D	A' (km²)	
Name of the main tributary R. Thames (the	$A':A$	
reservoirs are filled by pumping)		

	I	II
Average inflow m³/sec.	0.92	NIL (standing reserve)
Average outflow m³/sec.	0.92	NIL
Theoretical retention time	59 days	infinite

Geological characteristics

Thames valley alluvial gravels underlain by Oxford clay

Climatic conditions: Temperate maritime

Average monthly temperature max 7.5, 5.2, 4.8, 14.0, 14.5, 19.9, 19.8, 20.6, 17.8, 15.4, 9.3, 5.5,
min 1.2, 1.2, 1.9, 2.6, 5.2, 9.8, 10.1, 17.6, 9.1, 9.5, 4.2, 0.4

Average precipitation/year 833 mm Ice cover (days) 0–5

Average sunshine duration not known Average radiation/year not available

Main wind direction(s) SW % of calm days 2

Evaporation per year not known

Cultural geography and demography: On eutrophic river below one major sewage outfall and mainly agricultural land

Land usage of catchment area (%)   Usage of lake water including
                                   recreation activities
Industrial 5   Water supply
Agricultural 5   Sport fishing
Meadows 85   Bird watching (wild-fowl)
Forest 5   Sailing
unused 0
Number of residents none   inhabitants/km² not known
Water temperature: min < 1°C   max 23°C
Secchi depths: min 0.75 m   max 10 m
Euphotic zone: min 2.0 m   max total column
O₂ concentration: min 80%   max > 150%

pH 8.4   Conductivity (μS) I 600   Alkalinity I 200 mg · l^{-1}
                                II 500   II 100 mg · l^{-1} (as CaCo₃)

Average P-conc. I 200 μg · l^{-1} Average N-conc. I 3.0 mg · l^{-1}
                 II <20 μg · l^{-1} II 2.0 mg · l^{-1}

Conditions of sediment:
  I Reservoir completed 1965. About 4 cm of fresh sediment.
  II Reservoir completed 1977. About 1 cm of fresh sediment.

Dom. phytoplankton species:
  I and II Stephanodiscus astraea and hantzschii, Asterionella, Oocystis and other colonial green algae, Microcystis

Dom. zooplankton species:
  I and II Daphnia hyalina var. lacustris

Dom. macrophytes:
  I and II None

Dom. benthic organisms:
  I and II Chironomids, snails

Fishes:
  I and II Stocked with trout (Rainbow), coarse fish removed by irregular netting

4.2 Botanical Aspects

SEASONAL VARIATION OF PHYTOPLANKTON BIOMASS AND PHOTOSYNTHESIS IN THE HIGH-MOUNTAIN LAKE LA CALDERA (SIERRA NEVADA, SPAIN)

R. MARTÍNEZ

Abstract

Studies on lake La Caldera (Sierra Nevada, Spain) were carried out to follow changes in nutrients (N and P), phytoplankton biomass, photosynthesis and chloropyll a content. Differences were observed between two periods, 1974 and 1976. An increase in the PO_4 content of the lake occurred which could not be satisfactorily explained. Phytoplankton biomass was as much as twice (maximal values of 500 and 1000 mg m^{-3}, respectively) during the second year as the first. Photosynthesis also increased, although less than biomass; values near the maximum (2.5 mg C m^{-3} h^{-1}) at midday) were more frequent and production was more uniformly distributed in the second year. Biomass and production increases are attributable to PO_4 increase. Chlorophyll content per unit biomass varied between 0.3 and 1.6%, the lowest values belonging to the beginning and the end of the period. The relationship C ass h^{-1}/mg Chl a for each depth varied between zero (no production, under ice cover) and 1.1, 50% of the quotient values falling between 0.02 and 0.2, these low values being characteristic for an oligotrophic lake.

Introduction

La Caldera is a small, oligotrophic high-mountain lake situated in the Sierra Nevada (Granada, South Spain), at 3,040 m. a.s.l. Some of its basic features have been studied elsewhere (Martínez, 1975; Martínez, 1977). This paper is concerned with phytoplankton biomass and photosynthesis and their temporal variation.

Material and methods

Nitrate and phosphate were analyzed by spectrophotometric methods (Golterman, 1969).

Phytoplankton species, numbers and biomass were determined by taking 100 ml-water samples from the lake at the following depths: 0, 1, 2, 4, 6, 8 and 10 m (the last, whenever possible). Samples were sedimented for periods of 48 hours and examined under an inverted microscope. After determining species and counting specimens, biomass, as fresh weight, was estimated by taking the volumes of known geometrical bodies similar to the various algal species, assuming the specific weight to be 1.

Primary production was measured by calculating C assimilation in samples labelled with ^{14}C and incubated "in situ" for 4 hours from 10 to 14 hours, solartime, at the above mentioned depths. Then, they were filtered through cellulose nitrate filters of 0.2 μm pore diameter. A water pump was used, at low pressure to prevent any cell breakout. Filters were washed with 0.01 H NCl to eliminate possible residues of non-assimilated labelled bicarbonate onto the cells or the sestonic material. Filters were dried at room temperature. Dry filters were introduced in vials containing about 8 g l^{-1} Fluoralloy (a mixture of PPO and methyl PBD), dry mixture. ^{14}C counts were done by liquid scintillation. Total CO_2 was determined from pH and alkalinity values, its values ranging from 2.4 to 5 mg C l^{-1}. No corrections were introduced for different assimilation rates of ^{14}C and ^{12}C.

Chlorophyll a was estimated by spectrophotometry, according to Talling & Driver (1963), from acetone extracts of samples of 2 to 4 litres taken at each sampling depth and filtered through GF/C glass fiber filters.

Results and discussion

The study was carried out in 1974, from the beginning of July, at the start of the ice break-up period, until the middle of December, when there was 1 m of ice cover, and also in the summer of 1976, from the beginning to the end of the ice-free period (July to October)

Nutrients

In the summer of 1973, no PO_4-P had been detected. In the summer of 1974, it appeared only in some isolated samples and never exceeded concentrations of about $0.5 \mu g l^{-1}$. In the summer of 1975 phosphate in appreciable amounts began to appear, ranging from undetectable to $1.5 \mu g l^{-1}$.

In 1976, phosphate concentration at the ice-breaking time when sampling started (mid-July) was about $1.3 \mu g l^{-1}$ between 0 and 4 m depth. Previous higher concentrations are to be supposed, for the algal biomass was already relatively high by that time.

In Fig. 1 the depth-time distributions of phosphate and nitrate during the summer of 1976 are shown. At the end of July, when the biomass had reached its maximum value and phosphate was undetectable, the nitrate had decreased to a concentration as low as $6 \mu g l^{-1}$. This situation had not been detected before as, due to the phosphate limitation, the nitrate had not been so fully utilized by the algae. Subsequently the concentrations of nitrate rose to around $150 \mu g l^{-1}$, and then slowly decreased to about $50 \mu g l^{-1}$ by the end of the summer. This was probably due to an algal decay following the nutrient-limitation, with nutrients then being recirculated and allowing new algal growth to take place. PO_4-P reached concentrations between 4 and $20 \mu g l^{-1}$ and NO_3-N, between 54 and $67 \mu g l^{-1}$. These findings do not agree with Pechlaner's statement (Pechlaner, 1971) that, owing to the low water temperature, PO_4 is not likely to be easily recycled in alpine lakes. The results obtained here suggest rather a quick mineralization in the summer conditions of

the lake (T. max. $= 13°C$). It is also necessary to take into account that La Caldera is not deep and there is not summer stagnation, so that nutrients can circulate more easily.

The increase of phosphate in the lake, in relation to previous years, may be partly due to the PO_4 content of the snow. In the years 1973 to 1975 precipitations were rather low (average monthly value of 141.9 litres m^{-2}), those years having constituted a "dry period", whereas in the winter 1975–76 snowfall was higher (average monthly value of $180 l m^{-2}$), and the water level after thawing was about 0.5 m. higher in 1976 than in 1974. As a consequence, phosphate entering the lake had to increase. The snowpack of lake La Caldera was not analyzed for nutrients, but it is known from the research of Barica & Armstrong (1971) that snow is the main factor contributing to the nutrient budget entering the ice cover of lakes, and data from Larsson show (1973) that there is about four times as much phosphate in the snow covering a lake as in the water beneath. The alternating of periods of higher and lower snowfall could result in a series of cycles of higher and lower productivity in mountain lakes that have snow, and rainfall, as the only water source.

Another factor involved could be cultural eutrophication, the amount of people visiting the lake for picnics being ever increasing. In this regard, we could be faced with a unidirectional shift to progressive eutrophication. The ultimate reason, however, for this appearance of such unusually high PO_4-concentrations in the lake during the summer remains unknown and much more research has to be done to clarify the phosphorus cycle in the lake, by estimating different P-forms, contents in the sediment, snow and ice, and surveying trends for several years.

Community composition

Figure 2 shows the percentage of each algal class, and its variation in time, in the total phytoplankton community. The community structure is basically the same in both years, as far as the main species are concerned, *Chromulina* and *Cyanarcus* being, succesively, the dominant genera. In December 1974, under ice, Chrysophyta again accounted for most of the biomass (which was very small by that time).

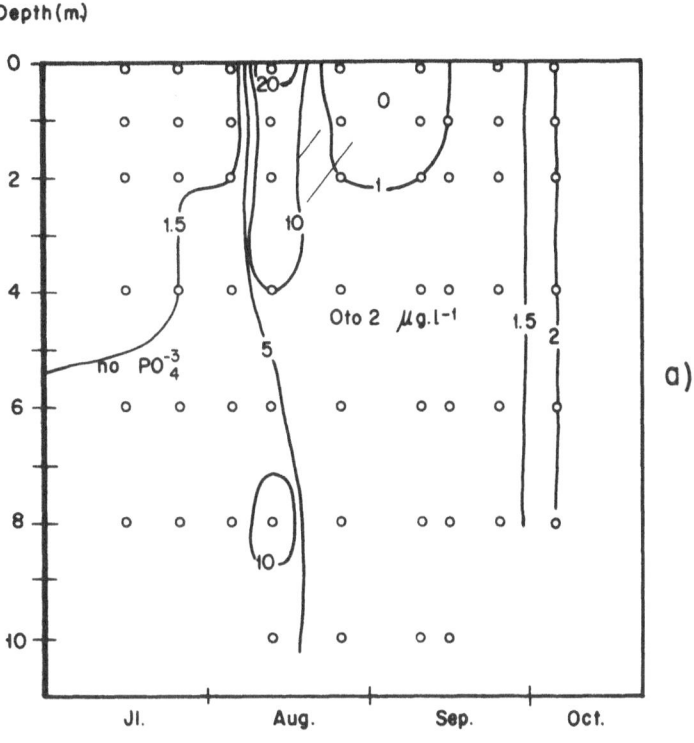

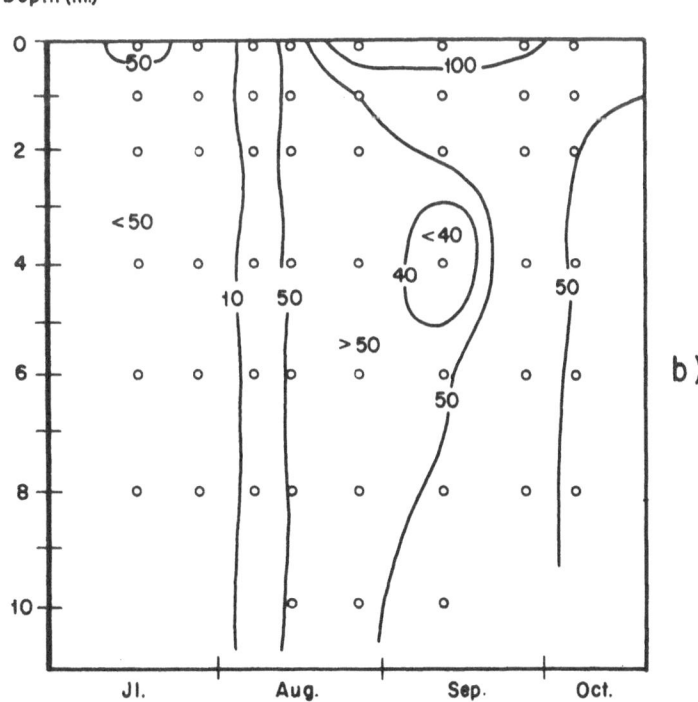

Fig. 1. Depth-time distribution of: (a) PO$_4$-P and (b) NO$_3$-N in lake La Caldera in 1976. Concentrations in μg l^{-1}.

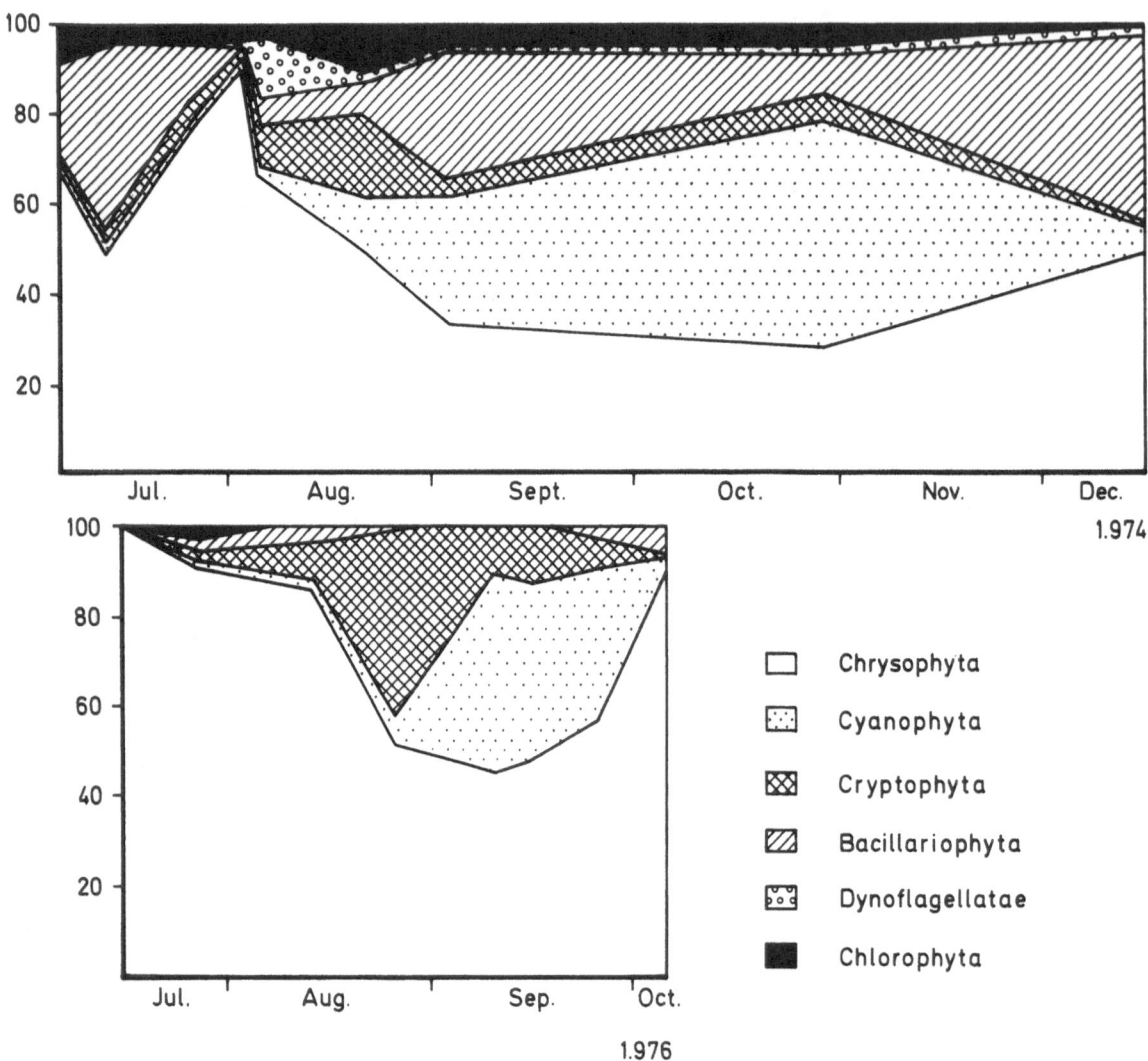

Fig. 2. Community structure of phytoplankton, and its variation in 1974 and 1976. Values are given as percentages of each algal class in the total biomass (fresh weight) per square metre.

Although biomass and primary production increased from 1974 to 1976, substantial changes, such as a shift to dominance by other algae (i.e. Chlorophyta) were not detected. This fact is in accordance with the observation made by Kalff *et al.* (1975) that in both an oligotrophic and a eutrophic arctic lakes (the latter having a much higher biomass) the community structure was very similar, made up mainly of Chrysophyta. As for high-mountain lakes, no obvious changes in species composition from year to year have been reported, with the exception of Kuoblatjakkojaure (Nauwerk, 1966).

In fact, in the summer of 1976 a slight simplification of the community structure was observed, the increase in biomass belonging mainly to the dominant organisms. Other algal classes present in small amounts (mainly Chlorophyta, Dinophyta and Diatomeae) were present in a lower proportion in the second year.

Biomass, photosynthesis and chlorophyll

Algal biomass reached a maximum value of $500 \, mg \cdot m^{-2}$ (fresh weight) at the beginning of August of 1974 and, in 1976, by the same period, a maximum of about $1,000 \, mg \cdot m^{-2}$. This trend to

114

increasing biomass was a consequence of increase of phosphate concentration in water.

Photosynthetic activity in the lake is low. Values for 1976 were higher than in 1974. Column maximum values were not much higher, but near-maximum values were reached more frequently. This is illustrated in Fig. 3, where column values (per m^2) of biomass and photosynthesis are shown.

The activity coefficient of phytoplankton per hour, at midday, (photosyntesis/biomass) was lower in 1976 than in 1974, for photosynthesis increased less than biomass. Mean values of this coefficient during the ice-free period were $2.1 \cdot 10^{-3} \cdot h^{-1}$ in 1974 and $1.1 \cdot 10^{-3} \cdot h^{-1}$ in 1976. For comparison purposes, zero P/B values estimated under ice-cover were not computed.

Figure 4 shows the distribution of photosynthesis according to time and depth in both years. Two main features should be emphasized:

First, photosynthesis was higher in the second year studied, values near to the maximum being more frequent.

Second, photosynthesis values were more evenly distributed, and not so stratified, either in time or depth, as they were in 1974.

Both data are in accordance with the statement made by Tilzer (1975) that an increase in concentration of a limiting nutrient (phosphate in La Caldera) would have a twofold effect: to increase the primary production and to reduce the light inhibition in water layers near the surface, as a consequence of the self-shading resulting from an increase in phytoplankton biomass. Phytoplankton of lake La Caldera seems to be light-inhibited at depths near the surface. Either a true inhibition or an avoidance behaviour could be postulated. In fact, both phenomena may take place. Figure 5 shows the depth-time distribution of the cell numbers of phytoplankton in both years studied. High cell numbers near the surface can be seen in some dates, which do not correlate with the photosynthesis rates (which are low) at the same depths.

Some phytoplankton profiles about the ice-break-up period (Martínez, 1977) showed that phytoplankton accumulated near the bottom. But this phenomenon seemed to be rather due to the higher nutrient concentration close to the bottom, before water circulation and mixture could take place (Pechlaner, 1971) than to an avoidance be-

haviour. Anyway, phytoplankton densities near the surface are often lower than at middle depths, which can be due to an avoidance behaviour that is occurring to a limited extent. For phytoplankton vertical profiles which are rather uniform photosynthesis continues being lower at depths near the surface, suggesting light-inhibition in the lake. In these conditions, the mentioned twofold effect can take place.

Concentrations of chlorophyll a ranged between $0.034 \, mg \, m^{-3}$ and $0.34 \, mg \, m^{-3}$. Column values ranged between 0.7 and 1.13 mg chl. $a \cdot m^{-2}$. The highest concentrations corresponded to the middle of July, resulting in an increase in C-assimilation by the end of July/beginning of August. The mean percentage of chlorophyll a in the water column ranged from 0.1 to 1.6% of the biomass (fresh weight), the highest value also corresponding to the middle of July, when chlorophyll showed the maximum concentration in both years. This data suggests that the proportion of chlorophyll a in the biomass is rather variable. Rott also found similar results (Rott, 1975) in his study of Piburger See (Tyrol), with the percentage of chlorophyll in biomass varying with the time of the day and the period of the year. This variability has to be taken into account when taking chlorophyll as an equivalent to biomass.

The quotient photosynthesis/chorophyll a, at midday, varied between zero (inactivity under ice) and 1.12 in both years. Higher values corresponded to dates when biomass was low but showed a great photosynthetic activity (end of July). With increasing biomass the ratio decreased. This phenomenon has been observed by Tilzer (1978) for the "photosynthetic capacity" of phytoplankton (value of the quotient at light optimum) for oligotrophic high-mountain lakes, and by Margalef et al. (1976), for the quotient value at midday, for other kinds of freshwater bodies. In lake La Caldera, 50% of the quotient values fell between 0.02 and 0.2, as belongs to the oligotrophic character of the lake.

Photosynthesis in lakes during the ice-covered period has been reported by several authors (Pechlaner 1971; Rott 1975; Rigler 1978, and others), but this was not the case in lake La Caldera in 1974, at any depth. Obviously, more experimental research needs to be made during

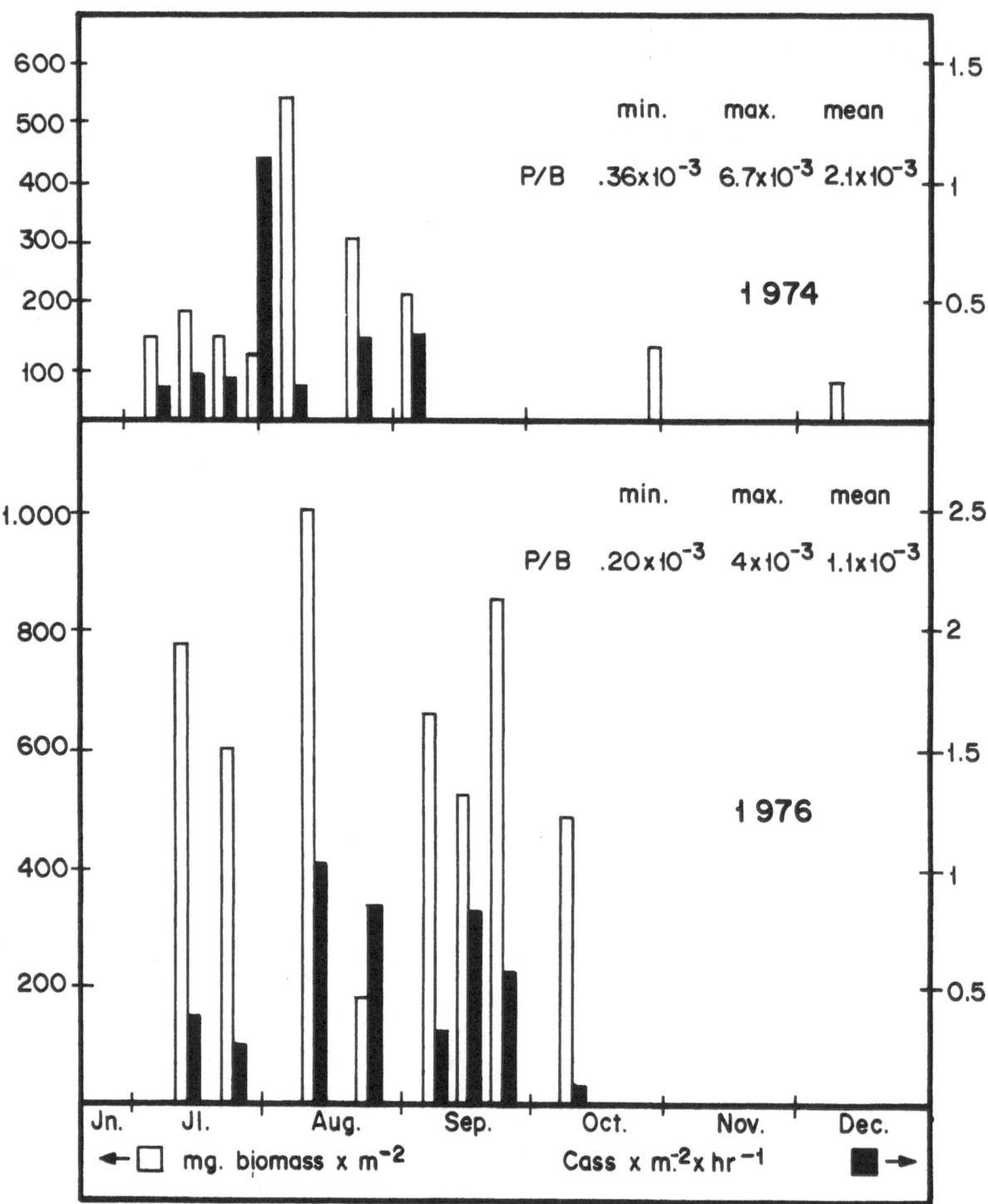

Fig. 3. Total column values of phytoplankton biomass and photosynthesis in 1974 and 1976. P/B ratios are in mg Ch^{-1}/mg biomass. All the photosynthesis values correspond to short term exposures around midday (all the incubations made from 10 to 14 hours).

116

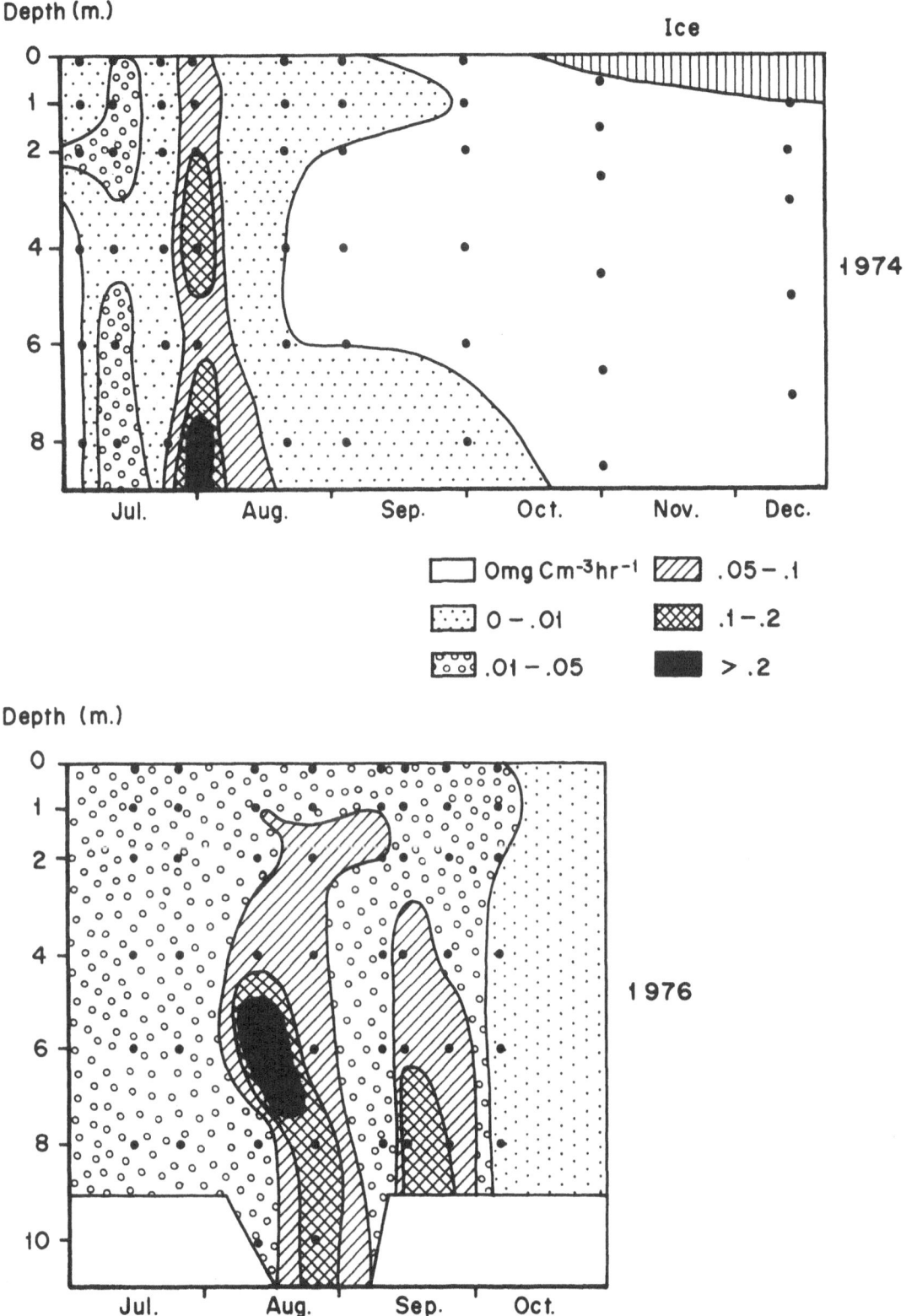

Fig. 4. Depth-time distribution of photosynthesis in lake La Caldera. All the values correspond to short-term exposures, made from 10 to 14 hours.

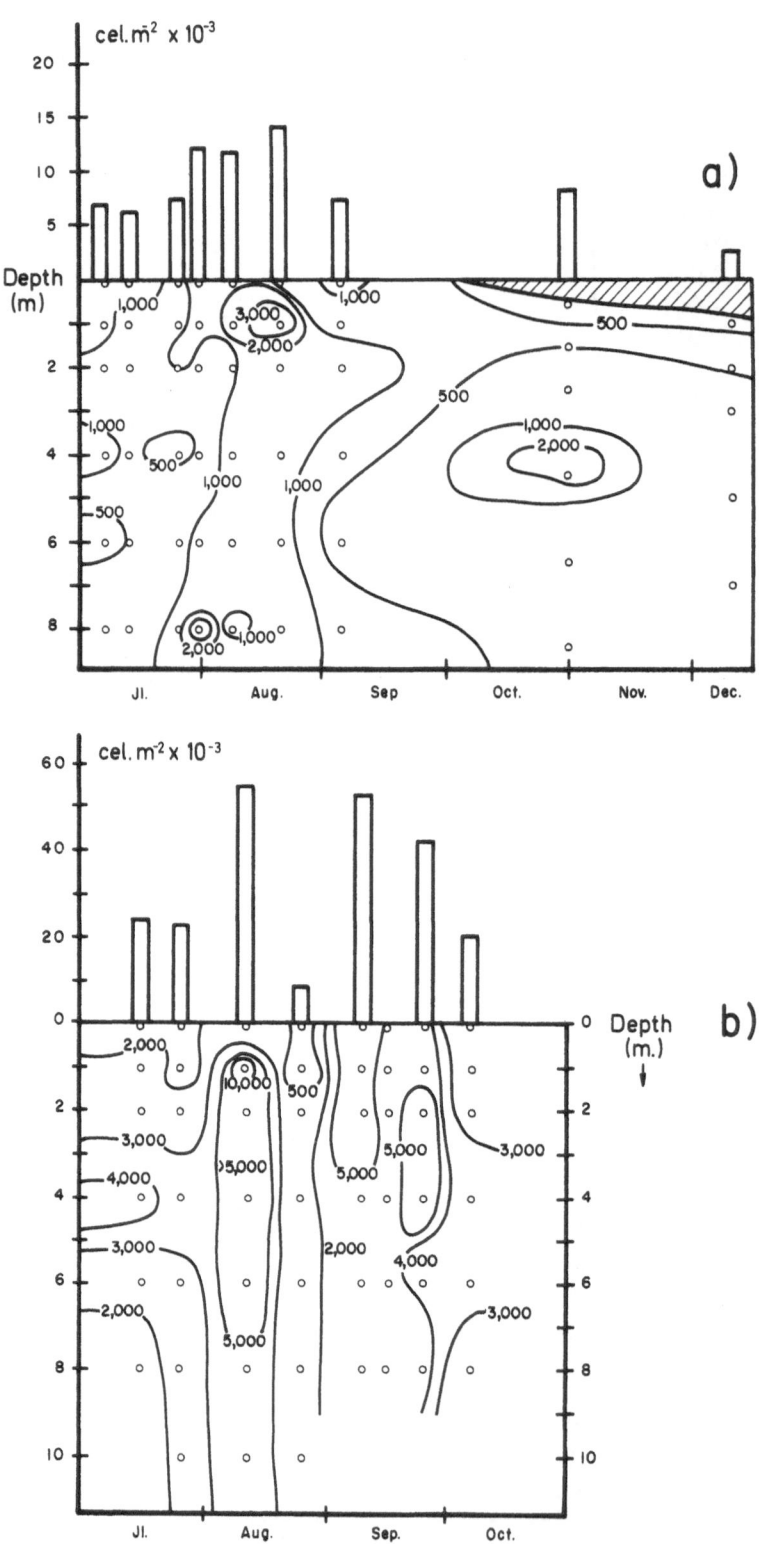

Fig. 5. Depth-time distribution of the phytoplankton cell numbers: (a) in 1974; (b) in 1976. Values are given in cells per cubic meter. Column values have been multiplied by 10^{-3}.

118

ice-cover periods, in order to produce a clearer picture of the production pattern of the lake.

Acknowledgements

This research has been partially supported by a grant of the Comisión de Investigación, Presidencia del Gobierno (Spain). I am indebted to Dr. D. Jewson for his kind criticism and suggestions on the manuscript. To the Service of Roads of the M.O.P. (Granada) I am most grateful for transportation facilities into the lake's area.

References

Barica, J. & Armstrong, F. A. J. 1971. Contribution by snow to the nutrient budget of some small northwest Ontario lakes. Limnol. Oceanogr. 16: 891–899.

Kalff, J., Kling, H. J., Holmgren, S. H. & Welch, H. E. 1975. Phytoplankton growth and biomass cycles in an unpolluted and in a polluted polar lake. Verh. int. Ver. Limnol. 19: 487–495.

Golterman, H. L. (ed.) 1969. Methods for Chemical Analysis of Fresh Waters. Blackwell, Oxford.

Larson, G. R. 1973. A limnological study of a high-mountain lake in Mount Rainer National Park, Washington State, U.S.A. Arch. Hydrobiol. 72: 10–48.

Margalef, R., Planas, D., Armengol, J., Guiset, A., Toja, J. & Estrada, M. 1976. Limnologia de los embalses españoles. 422 pp. Ministerio de Obras Públicas, Madrid.

Martínez, R. 1975. First report on the limnology of the alpine lake La Caldera in the Penibetic Mountains (Sierra Nevada, Granada, Spain). Verh. int. Ver. Limnol. 19: 1133–1139.

Martínez, R. 1977. Phytoplankton species, biomass and diversity in lake La Caldera (Sierra Nevada, Spain). Acta Hydrobiol. 19, 95–107.

Nauwerk, A. 1966. Beobachtungen über das Phytoplankton klarer Hochgebirgseen. Schweiz. Z. Hydrol. 28: 4–28.

Pechlaner, R. 1971. Factors that control the production rates and biomass in high-mountain lakes. Mitt. int. Ver. Limol. 125–145.

Rigler, F. H. 1978. Limnology in the high Arctic: a case study of Char Lake. Verh. int. Ver. Limnol. 20: 127–140.

Rott, E. 1975. Phytoplankton Biomass, Primärproduction und kurzweltige Strahlung im Piburger See. Doktor. Thes. Universität Innsbruck.

Talling, J. F. & Driver, D. 1961. Some problems in the estimation of chlorophyll a in phytoplankton. Univ. Hawaii. Atomic Energy Com. Publ. TD-7633. Ed. M. S. Doty.

Tilzer, M., Goldman, Ch. & De Amezaga, E. 1975. The efficiency of photosynthetic light energy utilization by lake phytoplankton. Verh. int. Ver. Limnol. 19: 800–807.

Tilzer, M., Goldman, Ch. & De Amezaga, E. 1978. Predictions of productivity changes in Lake Tahoe at increasing phytoplankton biomass. Ver. int. Ver. Limnol 20: 407–420.

PHYTOPLANKTON IN 35 FINNISH BROWN-WATER LAKES OF DIFFERENT TROPHIC STATUS

V. ILMAVIRTA

Abstract

Phytoplanktonic biomass, species composition, and the main chemical parameters of the water were measured during summer stratification in 35 humic lakes in Finland. The lakes were situated in four chains with differing tropic status. Green algae were dominant in eutrophic lakes, blue-greens in mesotrophic, but cryptophytes and chrysophytes in oligotrophic brown waters. Diatoms were abundant in all types. The ecology of flagellated species, found significant for brown-water lakes, is discussed in relation to allochthonous loading of the lakes.

Introduction

Typical of Finnish lakes is their dark-brown water colour. According to the data of the National Board of Waters (Laaksonen, 1970) the brownest waters ($\bar{x}$ 180 mg Pt l^{-1}) occur in the rivers on the west coast of Finland in the land up-lift clay areas. Here the water discharges from the Suomenselkä watershed which is covered in dense coniferous forests. Most of the clear-water lakes are in northern Finland ($\bar{x}$ 18 mg Pt l^{-1}). However, in many areas of central and southern Finland ($\bar{x} \sim 80$ mg Pt l^{-1}) the water of small forest lakes can be extremely brown, even over 450 mg Pt l^{-1}.

The chemical composition of Finnish lakes is quite well known, but only a few studies have been carried out on their phytoplankton communities and most of these are based on one particular lake. It is not known which phytoplankton species are important or what factors limit their productivity in different lake types.

This study compares the phytoplankton biomass and species composition of 35 Finnish lakes of different trophic status, nutrient concentrations, and water colour. Classification of these lakes, based on extensive studies of macrophyte vegetation, are in progress (unpubl. data of Toivonen, see Toivonen 1980). The phytoplankton and water chemistry of these lakes will also be used for classification. The main factors on which this will be based are described in this study and include phytoplanktonic biomass, and species composition, phosphorus and nitrogen concentrations, pH and transparency expressed as Secchi disc and water colour. Much attention has been paid to the dominant phytoplankton species, especially flagellates, which have been shown to be important for the production ecology of some humic lakes (cf. Ilmavirta, 1975, 1979, 1980). Most of the lakes in this study occur in chains, which is typical of Finnish lakes. The outflow from these chains discharges into a series of larger lakes and then into the Baltic Sea.

The lakes in study

The 35 lakes in this study are situated in southern Finland (61°15'–61°30'N, 23°35'–24°10'E) between the large lakes of Roine and Pyhäjärvi, both belonging to the water course of the River Kokemäenjoki. The lakes form four chains, two discharging into Lake Roine and two into Lake Pyhäjärvi. Six of the studied lakes are outside these chains.

Chain A (5 lakes) receives nutrient rich water from agricultural lands as well as large waste water inputs from settlements. It is highly eutrophic: algal blooms during summer stratification are abundant. Chain B (6 lakes) receives domestic sewage from some small settlements, but it is also heavily affected by car traffic along its shore. Chain C (10 lakes) discharges from dense coniferous forests and then through an extensive agricultural area and is largely oligotrophic. Chain D (8 lakes) is mostly located inside dense coniferous forests; these lakes are highly oligotrophic and only very slightly affected by forestry.

The total number of lakes in the study area (105 km²) is something over 200, a typical figure for these areas of Finland.

Most of the lakes in the study are small and shallow (mean and range):

	Area (ha)	Maximum depth (m)
Chain A	43 (9–156)	2.5 (1.6–3.5)
Chain B	56 (1–145)	3.8 (2.0–5.5)
Chain C	6 (0.5–14)	4.1 (1.5–8.5)
Chain D	10 (1–26)	5.5 (2.0–12.5)

The residence time of water is short (about 3–5 weeks; exact values not known).

Methods

The lakes were sampled for phytoplankton and water chemistry at four times: (I) during the overturn period in spring 1977 (24–28 May), (II) during strong thermal stratification in the summer of 1977 (25–29 July) and (III) 1978 (25–26 July), and also (IV) in late winter 1978 (5–9 April) when the lakes were still covered by ice. All the lakes were sampled within a period of 2–4 days: the time delay between the sampling of different lakes was then short enough to reduce any errors caused by rapid succession of phytoplankton after changes in climatic conditions.

One integrated sample of the epilimnion was taken in the deepest part of the lake for chemical analyses and phytoplankton studies. During stratification sampling was carried out at 0.5 m depth intervals through the whole epilimnion at 5–6 different horizontal stations. During the period of water circulation, if the temperature depth profile was even through the whole water column, only 2–3 depths were sampled. A plastic 1.0 l modified Ruttner sampler was used for sampling and a plastic container for mixing the samples. With this procedure it was possible to have the same water for analyses in all the bottles. This kind of mixed sample can be shown statistically to represent the open water area very well and overcomes the problems associated with vertical and horizontal spatial variations of water quality.

Samples for chemical analyses were filtered in the laboratory through Whatman GF/C glass fiber filters and stored untreated in the cold and dark. Phytoplankton samples were fixed in the field by an acetic acid Lugol solution and then two days later with buffered formaldehyde. Chemical analyses were done by Viljavuuspalvelu Ltd. using common standard methods (cf. Golterman et al., 1978). Phytoplankton were counted using inverted microscopy and a modified Utermöhl technique (see K. Ilmavirta & Kotimaa, 1974). Nomenclature is after Skuja (1948).

Because some part of the large phytoplankton material of this study is still untreated, only results sampled in period II, during the summer stratification of 1977, are discussed here.

I am sincerely indebted to Dr. Heikki Toivonen for his co-operation during the field work and for permission to use his unpublished data on macrophytic vegetation in these lakes. This work was financially supported by the National Research Council for Sciences (Luonnontieteellinen toimikunta) by the grant No. 179, which I sincerely acknowledge.

Results

In spite of their shallowness most of the lakes were thermally stratified during the summer period. The mean depth of the epilimnion in different chains was 2–3 m (range 1–5 m) with a mean temperature at 0.5 m depth of +16.5°C (13.5–17.3°C). The Secchi disc transparency of different chains was in A 0.8 m (range 0.45–1.4 m), B 1.7 m (0.25–3.2 m), C 1.3 m (0.7–3.0 m) and D 1.8 m (1.2–6.0 m).

Table 1. Some maximum and minimum values of measured chemical and biological parameters in different lake chains (A–D) during sampling periods I, II and III.

	range	min	max
pH	5.25 - 9.95	C	A
colour mg Pt/l	10 - 110	B	D
conductivity /uS$_{20}$°C	23 - 175	D	A
P tot /ug/l	5 - 420	D	A
PO$_4$-P /ug/l	2 - 320	D	A
N tot /ug/l	113 - 2460	D	A
NO$_3$-N /ug/l	1 - 1980	B	A
SiO$_2$ mg/l	0.07 - 9.44	B	A
Cl$^-$ mg/l	2.0 - 38.0	C	A
SO$_4^-$ mg/l	2 - 48	D	A
Fe mg/l	0.01 - 0.76	A	C
Inorg. C mg/l	0.5 - 15.6	D	A
Chl a mg/m^3	1.8 - 152.2	D	A
KMnO$_4$ cons. mg/l	8.8 - 65	A	C
Ignition loss mg/l	16 - 100	D	B
Partic. matter mg/l	20 - 160	D	A
N tot/ P tot	1:1 - 410:1	A	A
NO$_3$-N / PO$_4$-P	0.03:1 - 230:1	A	A
Phytoplanktonic prod. in vitro*mg C/m^3 a	12 - 3469	D	A

* Period III

Table 1, which includes all the material collected during ice-free periods (I–III), shows the great range of chemical lake types recorded in the study area. The clearest lake was found in chain B, the brownest in chain D. Highest nutrient concentrations were measured in chain A, lowest in D.

Some of the main chemical variables, expressed as a mean for each chain of lakes, are shown for period II in Fig. 1. Chain C and D, discharging from the forest area, were the darkest and most acid, but also extremely poor in nutrients. A and B showed high nutrient concentrations, conductivity and pH, with all these values being greater than the mean of Finnish lakes. NO$_3$-N concentrations were negligible during the stratification period, but 20–40% of the total phosphorus was as PO$_4$-P.

The mean nitrogen/phosphorus ratio was 38:1 for all the material sampled, which suggests a strong phosphorus limitation of phytoplanktonic production in all the lake chains if 7:1 is thought to be optimal. The highest N:P ratio, in this study, was 129:1 in Lake Vähä-Riutta of chain D, which was the only highly oligotrophic clear-water lake sampled, having groundwater seeping through the bottom sediments (Secchi disc 6 m, tot. P < 2 μg l^{-1}). The highest N:P ratio for brown-water lakes was 71:1 in chain A. Possible nitrogen limitation was found only in one lake, lake Kyläjärvi in chain A (N:P 3:1), which is heavily loaded by phosphorus from farms and artificially fertilized fields (tot. P 320 μg l^{-1}). Also the resuspension of sediments in this shallow lake (max. depth 3 m) affects the chemistry.

From the chemical values chain A could, overall, be classified as eutrophic with moderately

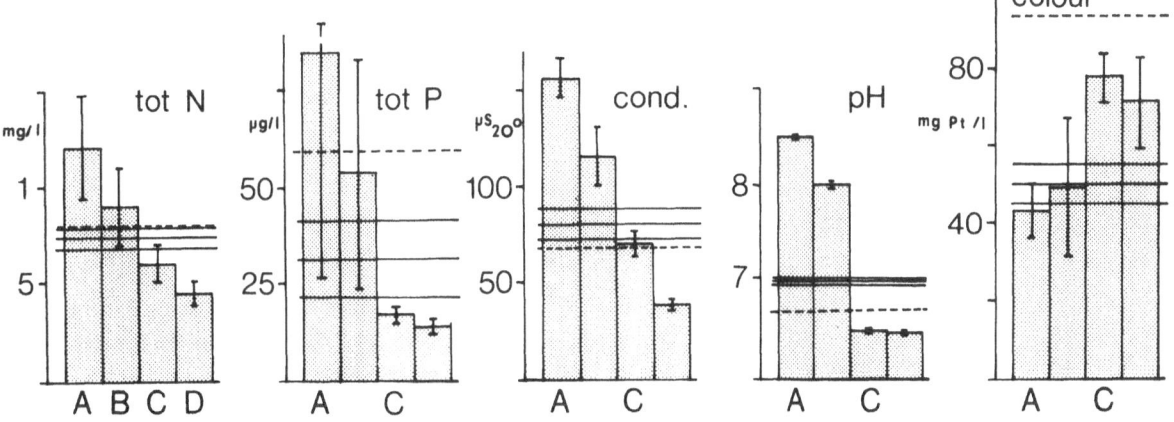

Fig. 1. The mean chemistry of water in lake chains A–D measured during the study period II. Vertical bars indicate the probable error of the mean, broken line the mean of Finnish lakes (Laaksonen 1969) and solid lines the mean with probable error of the mean in the whole material of the present study.

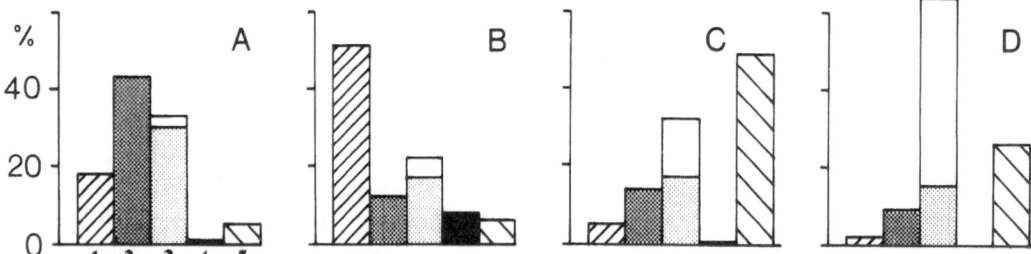

Fig. 2. The mean composition of phytoplankton in lake chains A–D as % from total biomass in the samples. 1 = Cyanophyta, 2 = Chlorophyta, 3 = Chrysophyta: lower part (shading) = Diatomeae, upper part = Chrysophyceae, 4 = Euglenophyta, 5 = Pyrrophyta (largely Cryptophyceae).

clear water, B as a mesotrophic clear water system, but C and D as highly oligotrophic brown-water systems. This very rough classification will be used in the following discussion on phytoplankton results. However, it must be noted that in some chains there are also lakes which do not conform very well with this classification.

The highest biomass 89.8 g m^{-3} fw was recorded from a lake in chain A, Lake Kirkkojärvi, which is heavily loaded by domestic sewage, and the lowest (112 mg m^{-3} fw) was in chain D from Lake Vähä-Riutta, earlier mentioned to be the most nutrient-poor lake. The mean phytoplanktonic biomass in different chains were as follows (range also given): A 33.0 (2.3–89.8) g m^{-3} fw, B 11.7 (1.1–39.3) g m^{-3} fw, C 2.8 (0.5–6.2) g m^{-3} fw and D 1.5 (0.1–3.7) g m^{-3} fw. Chlorophytes were the most notable group in the nutrient-rich chain A, their percentage decreased with decreases in nutrient concentrations and increases in brown water colour (Fig. 2). Blue-greens were most abundant in mesotrophic chain B but their proportion of the total biomass was negligible in nutrient-poor humic lakes. Diatoms had their maximum in the most eutrophic lakes, but the percentage of flagellated Chrysophyceae species increased significantly towards oligotrophic and more humic lakes. Flagellated cryptophytes were most abundant in the darkest lakes of chain C.

The biomass of different groups (g m^{-3} fw) is given in Fig. 3. Cyanophytes, chlorophytes and diatoms had higher biomass values only in the nutrient-rich lakes. Euglenophytes were most abundant in chain B (affected by domestic sewage), cryptophytes in A and C, but Chrysophyceae species had about equal biomass in all types of lakes.

The percentage contribution of biomass to different taxa in Fig. 3 shows that cyanophytes and chlorophytes were typical of the more nutrient-rich waters which have a low humus concentration, but chrysophytes and cryptophytes of oligotrophic humic lakes. Diatoms were significantly present in all types of lakes.

The total number of taxa (species and taxa without exact determination) was 169 in the studied samples. The highest numbers were found in the Chlorophyta (69) and the Chrysophyceae (34). Only the genus *Cryptomonas* and a group of small flagellates were found in all samples (frequency 100%). *Chroomonas acuta* and *Katablepharis ovalis* were lacking in two samples:

*Cryptomonas spp.	(Cryptophyceae)	100%
*Small flagellates	(Chrysophyceae)	100%
*Chroomonas acuta	(Cryptophyceae)	94%
*Katablepharis ovalis	(Cryptophyceae)	94%
*Dinobryon (8 species)	(Chrysophyceae)	>80%
*Chlamydomonas spp.	(Chlorophyta)	>80%
Oocystis (5 species)	(Chlorophyta)	>80%
*Uroglena americana	(Chrysophyceae)	>50%
*Mallomonas (4 species)	(Chrysophyceae)	>50%
Melosira distans	(Diatomeae)	>50%
Ankistrodesmus falcatus sl.	(Chlorophyta)	>50%
Crucigenia tetrapedia	(Chlorophyta)	>50%
Gloeocystis planctonica	(Chlorophyta)	>50%
Scenedesmus (12 species)	(Chlorophyta)	>50%

Most of these taxa are flagellates (asterisks in tabulation) belonging to the Chrysophyceae and Cryptophyceae but, although there is one diatom (*Melosira distans*) there are no blue-greens, euglenoids or Peridiniae species.

Merismopedia tenuissima, M. punctata, Aphanocapsa delicatissima and *Aphanothece chlatrata* are the only species in the Cyanophyta typical of oligotrophic waters. *Oscillatoria,*

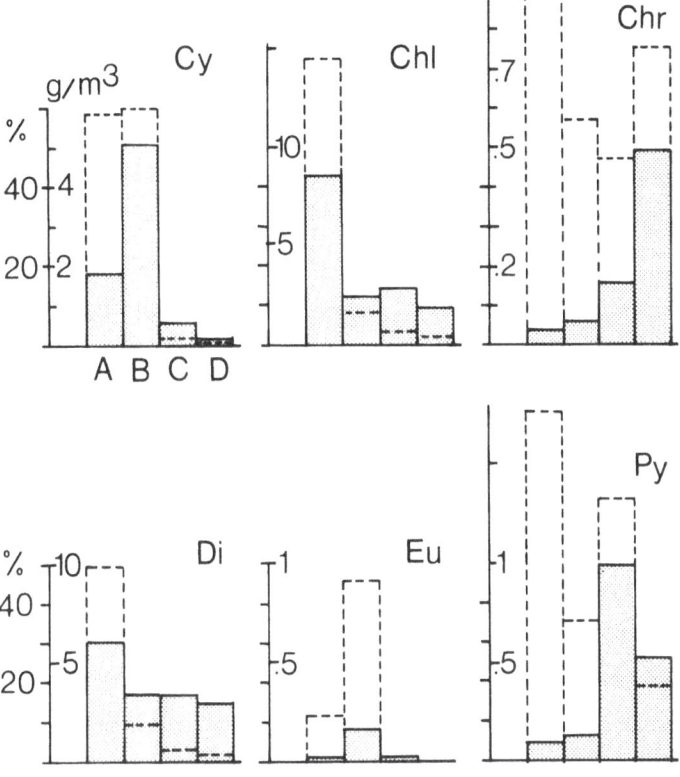

Fig. 3. The mean biomass (broken line) and the percentage biomass (solid line, hatching) of different phytoplanktonic groups in lake chains A–D.

Aphanizomenon, Anabaena and *Microcystis* species are well known indicators of eutrophy. Chlorophytes do not have any oligotrophic indicators in this material. *Ankistrodesmus falcatus* and most *Scenedesmus, Oocystis* (not *O. rhomboidea*) and *Chlamydomonas* species have their highest biomass in eutrophic moderately clear lakes. *Mallomonas akrokomos* and *M. allorgei* are clearly oligotrophic brown-water chrysophytes, but *Dinobryon* species and *Synura uvella* seem to be meso-oligotrophic. *Melosira italica* var. *tenuissima* is characteristic of highly eutrophic situations, but it may also grow to a high biomass in extremely dark oligotrophic lakes receiving large quantities of allochthonous material. *Melosira distans* is a good indicator of oligotrophy and *Synedra acus* of meso-eutrophy. Cryptophytes are quite indifferent, but their ecological amplitude is very broad. Euglenophytes were present in many eutrophic lakes, especially those receiving effluents from settlements. Also in some shallow lakes with the majority of the basin covered by submerged macrophytes some euglenophytes, especially *Trachelomonas volvocina*, can be abundant.

Discussion

A linear correlation between total phosphorus and phytoplanktonic production has been shown in Finnish humic lakes which receive domestic sewage but not in lakes affected by pulp mill effluents (Ilmavirta, 1979). This relation seems to hold if the colour of water is within the same order of magnitude (Granberg, pers. comm.; see also 1973) but if not then the light climate in the water seems to be the determining factor. Within the limits of nutrients and energy (light intensity, temperature) phytoplankton is regulating its seasonal succession in production, biomass and species composition (Ilmavirta, 1975, 1980; Jones & Ilmavirta, 1978).

In the present material only total nitrogen and total phytoplanktonic biomass showed a significant correlation ($r = +0.482$, $p < 0.1$) but the variation explained was poor ($r^2 = 23\%$). Chrysophyceae biomass also showed only a poor correlation to total phosporus ($r = +0.604$, $p < 0.01$, 36%). Correlations with other nutrients were even less significant as were also the correlations between nutrients and total number of species. These results suggest that there must be other factors involved, the most likely being light and temperature, although grazing and wash-out may also be important at times.

Diatoms were responsible for 94% and blue-greens for 72% of the variations in the total biomass, whereas cryptophytes and chrysophytes, although occurring in all samples, had little linear effect on the variations in biomass.

In the results from the lake chains it is possible to study the changes of the chemistry of the water running from lake to lake and to compare these results to the phytoplankton species composition and biomass. In Fig. 4 total phosphorus is plotted against phytoplankton biomass for lake chain C. This chain has two different legs which join and mix before entering lake 13. In the lakes of both legs there was a clear increase in phytoplanktonic biomass with increases in total phosphorus. However in lake 13, although the phosphorus concentration was higher, the biomas dropped and this decrease was further evident in the next lake in the chain (12), which was full of macrophytic vegetation dominated by *Typha angustifolia*. Nowadays the lake is more like a river, but still it is an effective nutrient filter. Phytoplankton (Fig. 5) collected near the outflow are similar to the community in Lake Pyhäjärvi (32), which is rep-

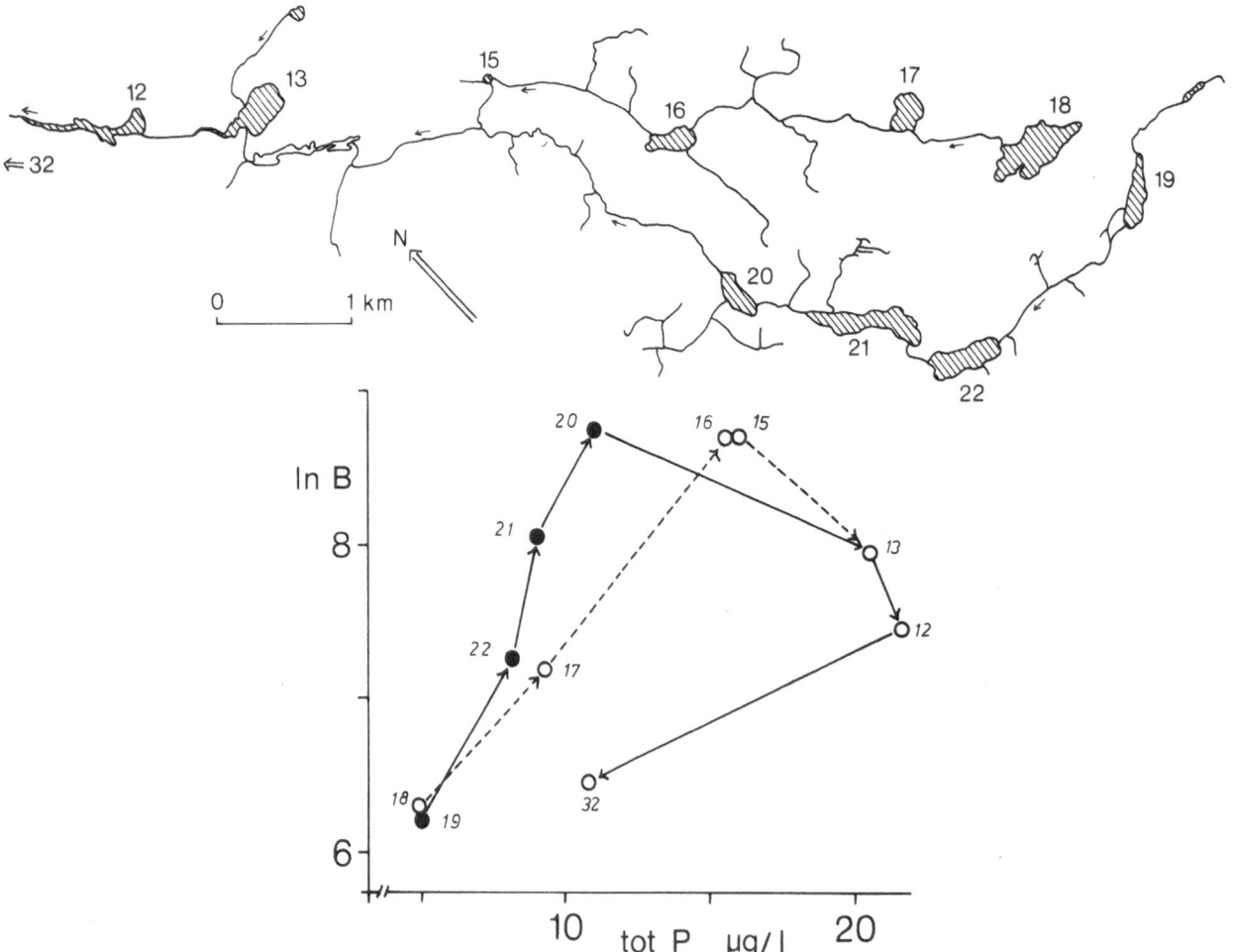

Fig. 4. Phytoplankton biomass (ln B. g m^{-3} fw) in lakes of chain C plotted against the total phosphorus concentrations in the water.

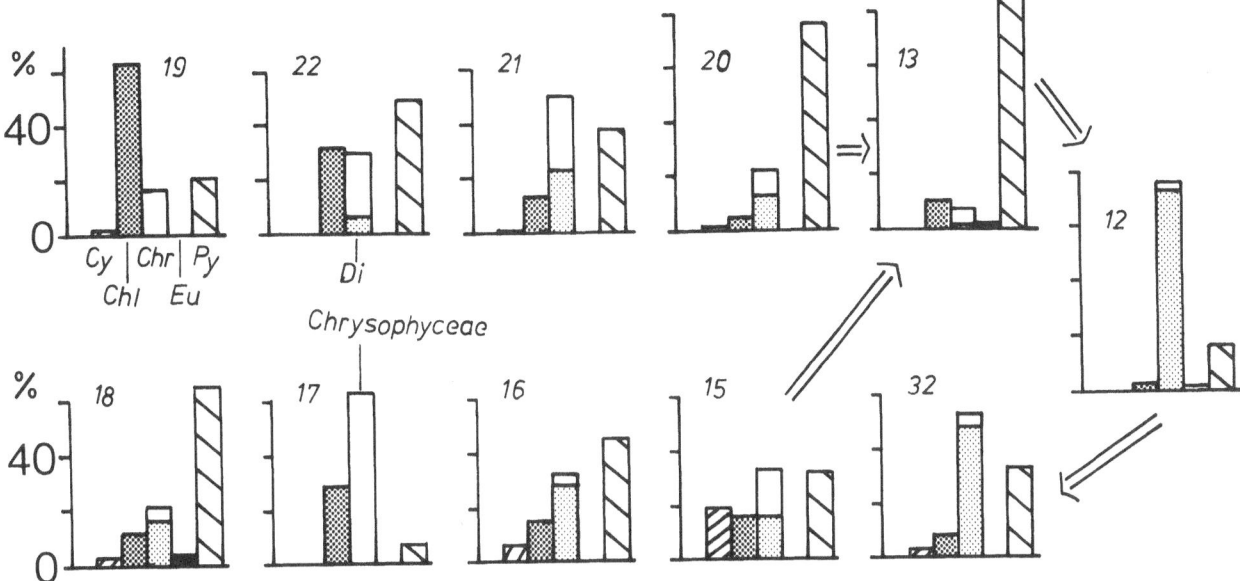

Fig. 5. The percentage contribution of phytoplanktonic biomass in different lakes of the chain C. For the numbering of lakes see Fig. 4.

resentative of the overall watershed. Compared to this lake the first lakes of both legs of chain C are really extremely oligotrophic.

There can be many theoretical explanations for the biomass drop in lake 13. The water might receive some toxic chemicals direct to the lake or in the river before it. Also some nutrients might have been used up before lake 13, thus limiting productivity. However, no evidence for those explanations can be found in our material. The most obvious reason for the drop of biomass is probably the mixing of two rather different phytoplankton communities in the same water body. In the northern leg of the lake chain (Fig. 5) the water was clearer ($\bar{x}$ 50 mg Pt l^{-1}) than in the southern leg ($\bar{x}$ 96 mg Pt l^{-1}), but conductivity higher ($\bar{x}$ 73 and 50 $\mu S_{20°C}$, respectively). In both legs flagellates were dominant, in the first lakes of southern leg chlorophytes were significant, but the biomass counts expressed as fresh weight were of the same magnitude (Figs. 4 and 5). Lake 15 is very small (0.5 ha) but relatively deep (max 4 m) and it is situated in a sheltered position in a small valley surrounded by forest. Thermal stratification in summer is strong in spite of short residence time and the hypolimnion is completely without oxygen. The phytoplankton community is characteristic of eutrophic conditions, with a high proportion of chlorophytes and cyanophytes. The mixing of this water with a quite different species composition from the brown water of the southern leg may result in a physiological stress on the phytoplankton. Species able to survive in a wide variety of conditions (= wide ecological amplitude) and with a better ability for competing in light or nutrient limiting conditions would be expected to predominate. The resulting situation in lake 13 was a domination by cryptophytes. In the present material cryptophytes had the widest occurrence in different lake types, they are also able to move actively to an optimal light intensity (Ilmavirta, 1974, 1975, 1980) and their small cell size ensures that nutrient uptake is rapid. All of these properties make the survival of cryptophytes especially favourable in humic lakes where there is a reduced penetration of light and/or low phosphorus concentrations (cf. Ilmavirta, 1975, 1980). Eloranta & Kettunen (1979) showed that *Cryptomonas* species are able to grow even in water strongly polluted by pulp mill effluents.

When plotting total nitrogen against phytoplankton biomass in chain C it was found that biomass was also less than expected in lakes 13 and 12. Chain B showed the same features (Fig. 6). One further explanation for the drop of biomass despite higher nutrient concentrations in

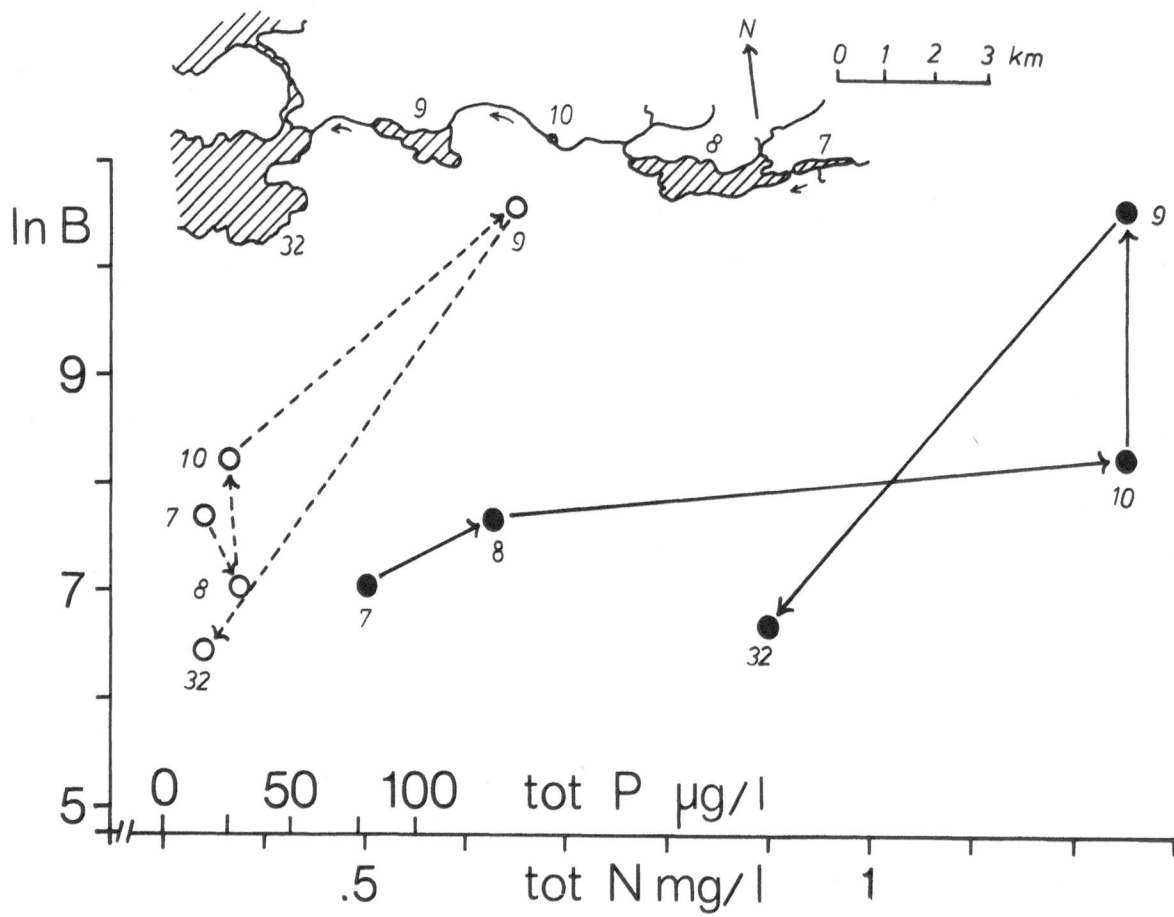

Fig. 6. Phytoplankton biomass (ln B, g m^{-3} fw) in lakes of the chain B plotted against total phosphorus (broken line, open circles) and total nitrogen (solid line, dots) concentrations in the water.

both chains could be the low capacity of brown-water ecosystems to eliminate external loading: organisms seem not to be able to convert nutrient loading to organic material after 20 μg l^{-1} P and 600 μg l^{-1} N in water. In Fig. 6 lakes 10 and 9 are heavily loaded by nitrogen, but the excess phosphorus in lake 9 evidently permitted nitrogen-fixing blue-greens (60% from total biomass) to manage. Lake 9 is badly eutrophicated and it can be seen visually in the rich development of macrophytic vegetation in this lake (cf. Toivonen & Ranta, 1976). The drop in biomass in chain C was also at 20 μg l^{-1} P and 600 μg l^{-1} N concentrations.

According to Toivonen (unpublished data) the effect of increasing nutrient load in chain C is seen also in aquatic vascular vegetation, since the

number of species of helophytes and hydrophytes increases continuously from lake to lake and their stands are more dense. Also the number of eutrophic species increases.

The role of flagellated species, especially of cryptophytes is significant in the production ecology of humic lakes. It has been shown in the brown-water Lake Pääjärvi in southern Finland (Ilmavirta, 1975, 1979, 1980) that the seasonal course of phytoplanktonic production is largely related to the water temperature and solar radiation, but not to the nutrient concentrations. On a seasonal scale there is a good correlation between phytoplanktonic biomass and productivity. The succession of total biomass is developed by the rapid continuous change in the phytoplankton species composition through the growing season and in this development flagellates are of major importance. In Lake Pääjärvi, during times of low turbulence, flagellates are able to migrate vertically following changes in light intensity and this seems to be one of the most important ecological adaptations in brown-water lakes. Also the change in pigmentation of phytoplankton have shown to be an important mechanism in light adaptation, whether at community or individual level (Jones & Ilmavirta, 1978).

A similar situation was also found in the present study: in dark lakes flagellates were dominant and often also had high biomass values. Theoretically it may be that in brown-water lakes flagellates are superior in their ecological tolerance, because they are able to move downwards during high light intensities at noon to avoid injury to their photosynthetic apparatus. In the morning and in the evening, or during overcast days, it is advisable to concentrate the cells at the surface to catch all the light available in dark water. Continuing with this hypothesis, in small stratified humic lakes, which mostly have a relatively shallow epilimnion (1–3 m, Table 2) and low turbulence, flagellates might sometimes migrate to the sharp boundary of thermocline to take up additional nutrients for their assimilation. The last idea is still without any experimental proof but for an algal cell that could take advantage of it there is a large potential store of nutrients available in the hypolimnion, evidently as a result of rapid bacterial activity in decomposition of allochthonous material, heavily loading most small humic forest lakes in Finland. This hypothesis is derived from the strong oxygen depletion in the hypolimnion of many lakes during summer stratification, in spite of the epilimnion being largely oligotrophic. Lake Iso-Riutta (No. 30) of chain D is a good example of this phenomenon: in the epilimnion (0–2.5 m) pH was 6.4, conductivity $36 \mu S_{20°C}$, tot P $10 \mu g l^{-1}$, tot N $400 \mu g l^{-1}$ and phytoplanktonic biomass 0.5 mg m^{-3} fw (89% flagellates), all these values suggest a highly oligotrophic status. Still at 12 m, 0.5 meter above the sediments the oxygen concentration was only 2 mg l^{-1}. In the epilimnion (0–1 m) of one of the most brown lake of this material (No. 28 Lake Helvetinjärvi = the devil's lake) pH was 6.1, conductivity $43 \mu S_{20°C}$, water colour 80 mg Pt l^{-1}, tot P $20 \mu g l^{-1}$, tot N $400 \mu g l^{-1}$ and biomass 3.7 g m^{-3} fw (95% flagellates), but in water from the hypolimnion there was a strong smell of hydrogen sulphide. These oxygen deficiencies must have resulted from an intensive bacterial decomposition of allochthonous material coming from the surrounding forests as these lakes are not meromictic (shown in winter

Table 2. Some hydrographical and chemical data on the lake water in chain C during period II.

| Lake | Max. depth m | Epilimnion | | | | | |
		Depth m	Temp. +°C	Secchi disc m	Colour mg Pt/l	Cond. $\mu S_{20°C}$	pH
18	8.5	?	16.5	>2	35	55	6.9
17	4.2	2.5	16.7	2.10	45	89	6.7
16	3.0	2.0	15.5	0.70	70	76	6.8
15	4.0	1.0	15.1	0.75	75	92	6.6
13	4.5	3.0	15.5	0.75	90	98	6.6
12	1.5	1.5	15.2	0.68	80	96	6.5
19	?	?	16.4	?	95	47	6.0
22	4.5	3.0	16.0	1.05	90	47	6.5
21	4.0	3.0	16.6	1.25	90	47	6.5
20	3.0	3.0	15.9	1.00	90	48	6.5

measurements) and phytoplankton biomasses are not high enough (algal bloom have not been detected).

To conclude the discussion it seems clear that the open water ecosystems of humic waters are very different than those described in clear water lakes. The role of humic material (the colloidal fraction causing the brown colour) and the particulate organic material loading this kind of lake is of much greater importance to the ecosystem than had been assumed before. In Lake Pääjärvi (Ilmavirta, 1979; Sarvala, 1978) the ecosystem receives about half of its primary energy as allochthonous material, but in small humic forest lakes the proportion of allochthonous material must be higher. In Lake Pääjärvi the rapid decomposition of organic material and succeding release of nutrients keeps phytoplanktonic production going. Thus the relation between decomposition by bacteria and grazing of zooplankton in releasing nutrient is of major importance for phytoplankton ecology. Therefore, when classifying lakes using phytoplankton the samples must be collected very rapidly (within a few days) to overcome the effects of rapid biomass succession resulting from different climatic conditions or grazing pressure.

References

Eloranta, P. & Kettunen, R. 1979. Phytoplankton in a water course polluted by sulphide cellulose factory. Ann. Bot. Fennici 16 (in press).

Golterman, H. L., Clymo, R. S. & Ohnstad, M. A. M. 1978. Methods of physical and chemical analysis of fresh waters. IBP Handbook 8. Blackwell Scientific Publications. Oxford, London, Edinburgh, Melbourne. 210 pp.

Granberg, K. 1973. The eutrophication and pollution of Lake Päijänne, Central Finland. Ann. Bot. Fennici 10: 267–308.

Ilmavirta, K. & Kotimaa, A.-L. 1974. Spatial and seasonal variations in the phytoplanktonic primary production and biomass in the oligotrophic lake Pääjärvi, southern Finland. Ann. Bot. Fennici 11: 112–120.

Ilmavirta, V. 1974. Diel periodicity in the phytoplankton community of the oligotrophic lake Pääjärvi, southern Finland. I. Phytoplanktonic primary production and related factors. Ann. Bot. Fennici 11: 136–177.

Ilmavirta, V. 1975. Dynamics of phytoplanktonic production in the oligotrophic lake Pääjärvi, southern Finland. Ann. Bot. Fennici 12: 45–54.

Ilmavirta, V. 1979. Sources and utilization of energy in Pääjärvi, an oligotrophic, brown-water lake in southern Finland. Arch. Hydrobiol. Beih. Ergebn. Limnol. 13: 212–224.

Ilmavirta, V. 1980. Phytoplankton vegetation, especially in humic lakes. Handbook of vegetation science. Dr. W. Junk b.v. Publisher, The Hague (in press).

Jones, R. I. & Ilmavirta, V. 1978. Vertical and seasonal variation of phytoplankton photosynthesis in a brown-water lake with winter ice cover. Freshwater Biology 8: 561–572.

Laaksonen, R. 1970. Vesistöjen veden laatu. Vesiensuojelun valvontaviranomaisten vuosina 1962–1968 suorittamaan tarkkailuun perustuva tutkimus (Water quality in the water systems. A study based on observations carried out by the water pollution control authority 1962–1968). Maa- ja vesiteknisiä Tutkimuksia 17: 1–132.

Sarvala, J. 1978. An ecological energy budget of the lake Pääjärvi. Lammi Notes 1: 12–16.

Skuja, H. 1948. Taxonomie des Phytoplanktons einiger Seen in Uppland, Schweden. Symbolae Bot. Upsaliensis 9 (3): 1–395, 39 Tafeln.

Toivonen, H. & Meriläinen, J. 1980. Impact of the muskrat, Ondatra zibethica. (L.), on aquatic vegetation in small Finnish lakes, (this volume).

Toivonen, H. & Ranta, P. 1976. Tampereen lidesjärven vesikasvistosta ja sen muutoksista. Changes in the aquatic vegetation and flora of the polluted Lake Iidesjärvi, SW Finland. Luonnon Tutkija 80: 129–138.

IMPACT OF THE MUSKRAT (*ONDATRA ZIBETHICA* L.) ON AQUATIC VEGETATION IN SMALL FINNISH LAKES

H. TOIVONEN & J. MERILÄINEN

Abstract

Impacts of the muskrat, *Ondatra zibethica* (L.), upon aquatic vegetation in a small shallow meso-eutrophic lake was studied by comparing recent vegetation to earlier data obtained from aerial photographs from 1941 onwards. Stands of *Phragmites australis*, *Schoenoplectus lacustris* and *Equisetum fluviatile* have diminished, whereas stands of *Typha latifolia* and transitory mixed communities of herbaceous helophytes have increased. Open water areas created by muskrats in depleted reed communities showed various short term successions, being first colonized by submerged and floating-leaved plants, and later by helophytes. Marked changes were also observed in some communities of floating-leaved plants.

Marked changes in aquatic vegetation due to the effects of muskrats were also observed when the recent evidence on dominant macrophytes was compared with data from the late 1940s in 38 small lakes located in another region of Southern Finland. Stands of some earlier dominants. especially *Schoenoplectus*, *Equisetum* and *Nymphaea* have considerably decreased or even become extinct, and have been replaced by transitory vegetation of other helophytes and floating-leaved species.

Introduction

The muskrat, *Ondatra zibethica* (L.) uses large quantities of aquatic macrophytes, both for food and housebuilding materials. Pelikan *et al.* (1971) estimated impacts of muskrats on stands of *Typha latifolia** in a well-vegetated Czechoslovakian lake, and found that an autumn population of

* Nomenclature for vascular plants follows mainly Ehrendorfer (1973) and that for mosses Koponen *et al.* (1977).

28–55 animals per hectare reduced the annual production of *Typha* by 5–10%. In a northern Swedish lake the areas of open water were estimated to represent a removal of about 4 percent of the *Equisetum* stands in the summer following a population peak of 3–6 animals per hectare (Danell, 1979). In places where there are higher densities of muskrats, due to either a low trapping intensity or to habitat change, e.g. after colonization, whole areas of vegetation may be eaten out. The effects have previously been reported from North America (cf. literature cited by Danell, 1977), but also from Europe (Artimo, 1960; Marcström, 1964; Akkermann, 1975).

Muskrats were introduced into Finland from North America and Central Europe mainly in 1920s and 1930s because of their valuable fur. It soon dispersed over wide areas probably becoming acclimatized by the early 1950s (Artimo, 1960). In Finnish waters this animal has been a successful colonizer, and has evidently occupied an empty niche due to lack of severe competitors or any greater pressure from predators. In such conditions muskrats can have marked impacts on aquatic vegetation, and some reports are already published (Brander, 1949, 1951; Artimo, 1960; Toivonen & Ranta, 1976; Meriläinen & Toivonen, 1979). Small well-vegetated lakes with relatively stable water levels are especially favoured by muskrats, but heavy impacts have been reported also in waters with poor stands of helophytes (Rintanen, 1976).

The aim of the present paper is to study the impact of muskrats upon the succession of aquatic vegetation during the last few decades, particularly in a small shallow lake (Keskimmäinen), and also in a larger group of small lakes in Southern Finland. The study on Lake Keskimmäinen was made jointly by the two authors (more details in Meriläinen & Toivonen, 1979), studies in the other lakes was made by the author Toivonen.

Study lakes and methods

The lake subjected to detailed study was the small lake Keskimmäinen (located in the Finnish Lake District, 62°53′ N, 28°15′ E). The area of the lake is 46 ha, and the shallow basin (max. depth 2 m) is wholly covered by aquatic vegetation (30 ha of reedy helophyte vegetation and 16 ha of hydrophyte communities).

The lake origin was by isolation about 200 years ago, when the water table of the greater water body, Lake Riistavesi was lowered by about 2 m. At the beginning of the present century the water level of Keskimmäinen was lowered again, by about 0.5 m. These events have caused rapid expansions of reedy helophytes, especially *Equisetum fluviatile*.

Drainage area of the lake is small, but the nearby esker area feeds the lake with ground water through numerous littoral and sublacustrine springs. Although the lake can be classified on the basis of water chemistry to mesotrophic (e.g. specific conductivity 80–90 $\mu S_{20°C}$ and pH varying 7–8 in the open water season), the constant input of nutrients from spring waters partly explains the occurrence of several hydrophytes preferring eutrophic waters.

Muskrats entered the lake Keskimmäinen in 1937–38 and reached a very dense population in a few years; in 1941 about 150 large winter houses were found (5 houses per hectare of reedy helophyte vegetation). Since then the animal has continuously had a strong impact on the vegetation. In the spring of 1979 about 50 conical winter houses were counted.

The lake was mapped with the aid of aerial photographs in 1959–60 (J. M.) and in 1976–77 (H. T. & J. M.). Supplementary data about the succession were also obtained from aerial photo-graphs from 1941 (taken on 3.VI. and thus showing remnants of helophyte vegetation from 1940, the year before the peak occurrence of the muskrat population), and in later years 1966 and 1973.

In order to evaluate observed vegetational changes in the Lake Keskimmäinen 38 small lakes (1–200 ha) in the surroundings of the city of Tampere (61°15′–61°30′N, 23°35′–24°10′E) were studied. These lakes represent a wide variation of different lake types (cf. also Ilmavirta, 1980). In these lakes old data from years 1947–50 about dominating helophyte and hydrophyte species (studied by the late Dr. U. Perttula, unpublished material) were compared with the reinvestigation in 1975–78.

Results and discussion

Lake Keskimmäinen

Shore meadows of the lake are dominated mainly by *Carex rostrata*. This zone was in 1959–60 succeeded outwards by extensive stands of *Phragmites australis*, *Equisetum fluviatile* and by some clones of *Schoenoplectus lacustris* (Fig. 1). *Typha latifolia* was rather sparse. Herbaceous helophytes grew in small stands or scattered at the margins of reed beds, or in places where vegetation was either damaged or in a transitory stage.

The hydrophyte vegetation was also luxuriant. Floating-leaved species *Potamogeton natans*, *Nuphar pumila* and *N. lutea* × *N. pumila* grew in large and dense stands. However, the main part of floating-leaved vegetation consisted of mixed stands of the forementioned species and *Nymphaea* spp., *Sparganium emersum*, *S. friesii* and *Sagittaria natans*. The most abundant elodeid was *Myriophyllum verticillatum*, however *Potamogeton berchtoldii*, *P. obtusifolius*, *P. perfoliatus* and *Sparganium minimum* were also frequent, although they grew mostly in mixed nymphaeid-elodeid vegetation. In pleustophyton eutrophic *Lemna minor*, *Ricca fluitans* and *Ricciocarpus natans*, eurytrophic *Utricularia vulgaris*, as well as pleustophytic aquatic mosses were abundant. Usually they grew mixed with nymphaeids and elodeids, but mosses *Calliergon megalophyllum*, *Drepanocladus tenuinervis* and *D. trichophyllus* could also form carpet-like stands, even 0.5–0.7 m in thickness, which almost excluded other submerged plants.

Changes in helophyte vegetation evident in 1976–77 are presented in Fig. 2. They partly represent the 'natural hydrosere succession' of the vegetation, which has been accelerated by two lowerings of the water table. The expansion of *Equisetum* stands in the middle and eastern parts of the lake (sites 14, 15, 19, 21 and 24) is mainly explained by these phenomena. Overall, the increase in the area of *Equisetum* stands has during the study period been 24%. Later these sheltered and shallow habitats may be expected to develop into reedswamp, mainly dominated by *Equisetum*, *Phragmites*, *Carex rostrata* and *Typha*.

Another main reason for the vegetation changes are the effects of muskrat housing and feeding (cf. Akkermann, 1975; Artimo, 1960; Danell, 1977, 1979; Pelikan *et al.*, 1971), which are most visible in the closed reedy helophyte stands. *Phragmites australis* (sites 1–4, 6, 8, 9, 11 and 12) and *Schoenoplectus lacustris* (3, 4, 5 and 23) as well as *Equisetum fluviatile* in the more open N part of the lake (sites 6–10) are closely reduced. In *Phragmites* the decrease in the total area of the stands from 1960 has been 21% and that of *Schoenoplectus* 86%. According to documentation of aerial photographs the decreasing trends have been continuous from 1941 onwards. However, *Typha latifolia* has since 1959 greatly increased its occurrences and is now one of the dominant species (sites 6, 18, 20, 23, 25–29). The invasion of *Typha* at the lakeward margins or in broken patches in reedy helophyte beds is greatly favoured by the destructions made by muskrats in closed stands of other reedy helophytes.

Instability of the recent vegetation due to the muskrats activities is also indicated by the agglomerate stands of herbaceous helophytes, which are more numerous than in 1959–60. This vegetation is developing in openings of broken reed beds, especially in shallow water. The most abundant species are *Cicuta virosa* and *Alisma plantago-aquatica*, but in addition smaller plants which are normally of only minor importance in the closed reedy helophyte vegetation also occur more frequently, e.g. *Calla palustris*, *Lycopus europaeus*, *Lysimachia thyrsiflora*, *Rorippa palustris*,

Epilobium palustre, *Bidens* spp. etc. (cf. Brander, 1951; Danell, 1977). Also floating rafts originating from broken fringes of the reed beds and rhizomes covered with low vegetation of small plants have become more numerous.

After destruction of reedy helophyte stands the recovery of vegetation depends on the exposure of the stands and the duration of muskrat feeding and housing at the site. At the more exposed lakeward margins (e.g. sites 1–7, 11) a pioneer vegetation consists mainly of elodeids. Later floating-leaved plants, *Sparganium* spp., *Potamogeton natans* and *Nuphar lutea* appear and form a more closed community, rich in elodeids and aquatic mosses. If the succession continues, without further interruption by muskrats, recolonization of reedy helophytes via open water and nymphaeid phases takes from 5 to 15 years. These successions, which involve the repetition of earlier successional stages of the plant community result in frequent changes in the dominant helophytes, and so to a greater diversity of the vegetation.

At sheltered sites the succession in the open water areas created by muskrats greatly depends on the earlier dominants of the stand. In *Equisetum* stands small openings (25–100 m^2) show a good recovery following the transitory vegetation of submerged plants and aquatic mosses. The site will normally be re-invaded in few years by vegetative spread by rhizomes (cf. Danell, 1977; 1979). At sites where greater areas are exposed and where rhizomes of *Equisetum* have died the recovery seems to be rather slow, and the succession is characterized at first by elodeid and pleustophyte vegetation. Later nymphaeids, agglomerates of herbids, *Typha* or *Equisetum* itself invade the habitat. In *Schoenoplectus* stands the destructive effects of muskrats many times lead to complete eradication, and the habitat will subsequently become occupied by other plants. In *Phragmites* stands a large number of broken shoots are normally left at the site, which makes the establishment of other plants more difficult. In Keskimmäinen, aquatic mosses, *Utricularia vulgaris*, *Potamogeton alpinus* and *Sparganium minimum* thrived well in the openings and thereafter according to the water depth either taller nymphaeids (*Sparganium emersum*, *Nuphar*

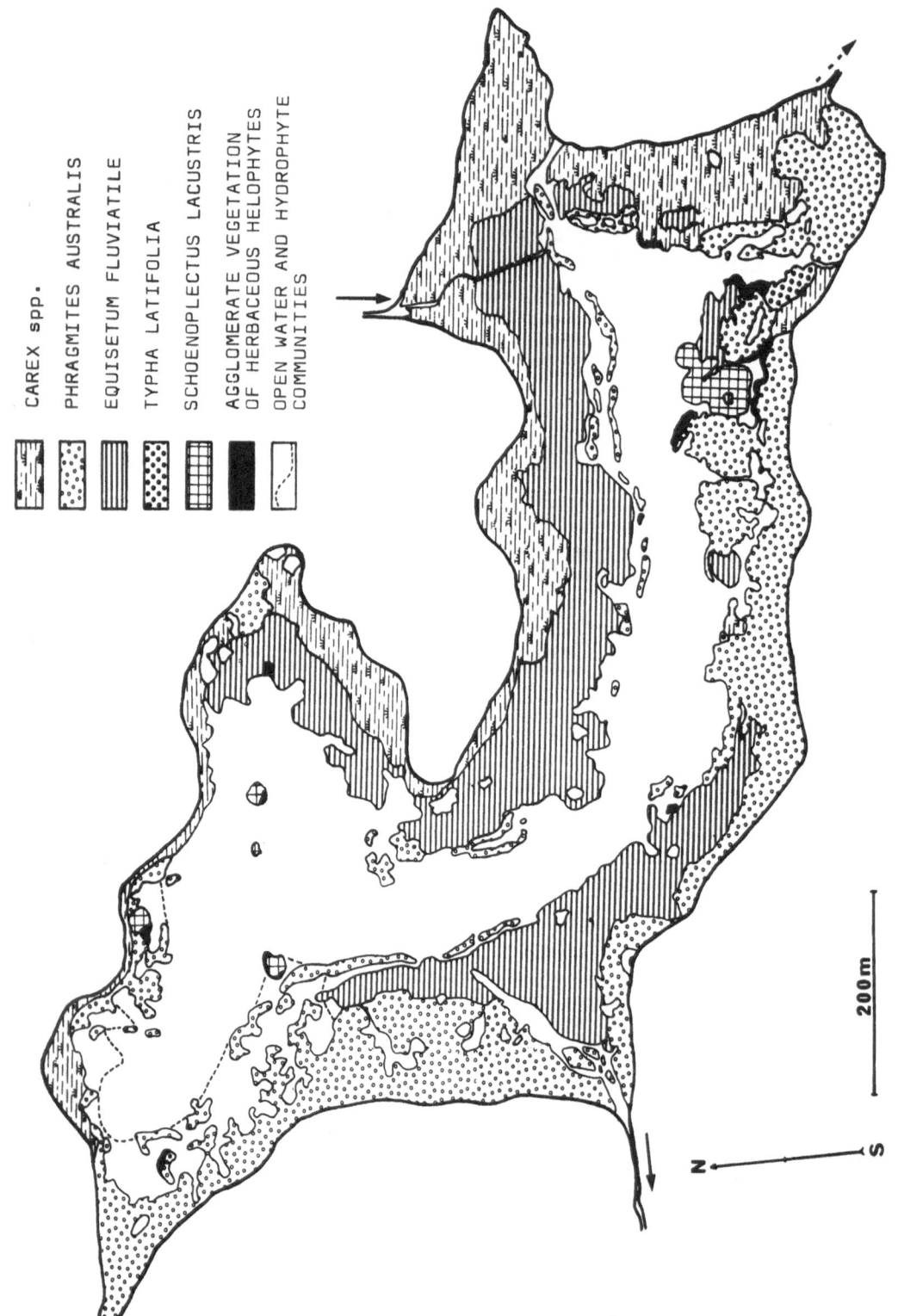

CAREX spp.

PHRAGMITES AUSTRALIS

EQUISETUM FLUVIATILE

TYPHA LATIFOLIA

SCHOENOPLECTUS LACUSTRIS

AGGLOMERATE VEGETATION
OF HERBACEOUS HELOPHYTES

OPEN WATER AND HYDROPHYTE
COMMUNITIES

200m

N

S

Fig. 1. The main helophyte stands (named according to dominant species) in the lake Keskimmäinen in 1960. The broken line indicates the lakeward margin of the helophyte vegetation in summer 1941.

134

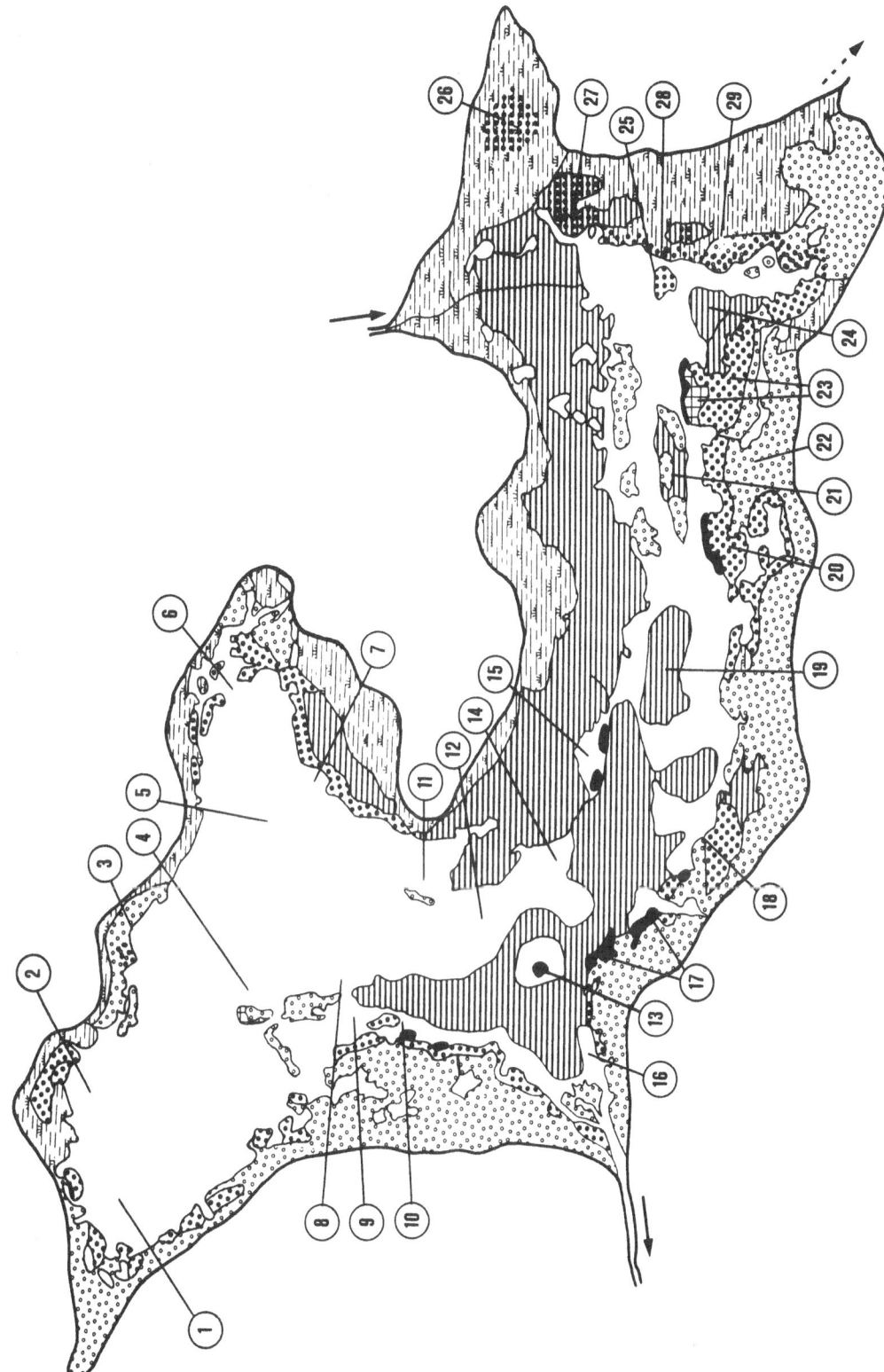

Fig. 2. The main helophyte stands in 1976. The numbers indicate the sites showing the most remarkable changes in the helophyte vegetation since 1960 (from Meriläinen & Toivonen 1979, Courtesy of Ann. Bot. Fennici).

lutea) or herbaceous helophytes, *Typha* and *Carex rostrata* entered the site.

The floating-leaved and submerged vegetation outside helophyte zones have also undergone considerable changes from 1960 to 1976. Alterations in abundance are mainly to be explained by feeding effects of muskrats and changes in the degree of shelter of habitats. The most remarkable change has been the decrease of *Nuphar pumila*, earlier dominant in the N part of the lake. The reason for this, as well as for the decrease of *Nymphaea tetragona* and *Sagittaria natans*, has been utilization by muskrats, especially of rhizomes during winter under icecover. In the S part *Nuphar pumila* and *Nymphaea tetragona* have presumably been replaced by taller *Nuphar lutea* and *Nymphaea candida* due to softening of the bottom caused by rapid sedimentation and increasing competition by elodeids and helophytes. In these areas the simultaneous effect of the hydrosere succession and activities of muskrats result in large transitory '*Equisetum–Sparganium–Potamogeton–Drepanocladus*'-communities.

Changes of dominant macrophytes due to muskrats in 38 small lakes

Comparison of recent dominant (according to the total area of dense continuous stands) macrophytes with earlier dominants in 38 small lakes is given in Table 1. In order to study the effects of muskrats this kind of rough comparison of changes in vegetation must be treated with extreme caution to differentiate it from the many kinds of successions caused by different environmental factors and human influences (lowering of water table, eutrophication, decrease of grazing by domestic animals), which may be coincident.

Some of the trends observed in Lake Keskimmäinen were also evident in changes of dominating helophytes and nymphaeids in the lake group studied. *Equisetum fluviatile* and *Schoenoplectus lacustris* have decreased markedly, and in some lakes their populations have been completely depleted by muskrats. *Schoenoplectus* seems to persist only in deep-water locations (cf. Danell, 1978). Also recovery of stands of these species has been quite effective. Instead, communities of other reedy helophytes or floating-leaved plants have developed at the sites.

Populations of *Phragmites australis* are now of about the same quantity or they have slightly increased. This is the result of not only the greater ability of this species to withstand muskrat feeding compared to the above mentioned species but also to the favourable effects of eutrophication (cf. Björk, 1967) and the decrease of grazing by domestic animals during the period of observations. *Typha latifolia* has increased in many lakes due to an increased instability of the vegetation. This species has especially invaded new sites and become more abundant in the slightly lowered lakes. In these (seven cases in this study) neither *Schoenoplectus lacustris* nor *Equisetum fluviatile* have invaded the new habitats created, although they are known to be effective colonizers (Aario, 1933); this is suggested to be due to the activities of muskrats as well.

In communities of floating-leaved plants *Nymphaea candida* has suffered and in some oligotrophic lakes their stands have been almost totally depleted by muskrats. Compared to *Nymphaea* it seems that *Nuphar lutea* is not strongly affected by muskrats; in this region possible harmful effects made by muskrats are wholly covered by the impact of increasing eutrophication, which greatly increases the density of the stands. The increase of *Sparganium* (*emersum*) and in some cases also of *Nuphar lutea* is mostly due to the slight lowering of the water table. In a few cases the destruction of the reed helophyte stands by the muskrats has favoured these species by creating new habitats (usually transitory) for them.

Changes in submerged plant communities due to muskrat activities do not become evident in this kind of analysis because of their subordinate role in the aquatic vegetation of these lakes. Only *Myriophyllum alterniflorum* is of interest in this connection. In some oligotrophic lakes where muskrats have destroyed the helophyte and nymphaeid vegetation severely, it subsequently becomes one of the dominants. Our field experience also suggests that muskrats do not normally have any considerable impact on submerged vegetation. In some cases, however, it has attained new temporary habitats after the destruction of closed helophyte or nymphaeid communities.

Table 1. Dominant macrophytes (according to the total area of dense continuous stands) in 38 small lakes in 1947–50 (I) and in 1975–78 (II). The number of dominants taken into study varies from 3 to 4 per lake, and is the same in the two surveys. The figures thus show the number of lakes where the species concerned occurred/occurs as a dominant. Figures in brackets show the number of cases where stands of the particular species have remained more or less unaltered, and its occurrence as the dominant is merely due to decline of earlier dominants. Eu- and mesotrophic lakes are usually densely vegetated, in dys- and oligotrophic lakes the vegetation cover is continuous, but not dense. In all lakes, however, vegetation cover has been dense enough to maintain muskrats at least for a short time.

Lake type	Eu- and meso-trophic		Dys- and oligo-trophic		All lakes	
	I	II	I	II	I	II
Dominant macrophytes						
Emergent:						
Phragmites australis	16	21(4)	4	6(2)	20	27(6)
Equisetum fluviatile	15	7	7	4	22	11
Schoenoplectus lacustris	11	3	–	–	11	3
Typha latifolia	1	10	–	–	1	10
T. angustifolia	1	1	–	–	1	1
Sagittaria sagittifolia	3	2	–	–	3	.2
Sparganium erectum	1	1	–	–	1	1
Butomus umbellatus	1	–	–	–	1	–
Iris pseudacorus	–	–	1	1	1	1
Floating-leaved:						
Nuphar lutea	18	20	15	15	33	35
Potamogeton natans	9	7	1	–	10	7
Nymphaea candida	2	2	5	1	7	3
Sparganium species	3	6(1)	2	8(6)	5	14(7)
Polygonum amphibium	1	1	–	–	1	1
Submerged:						
Myriophyllum alterniflorum	1	1	4	6(2)	5	7(2)
Elodea canadensis	1	2	–	–	1	2
Potamogeton obtusifolius	–	1	–	–	–	1
Sparganium minimum	1	1	1	–	2	1
Ceratophyllum demersum	1	–	–	–	1	–
Utricularia vulgaris	–	–	3	2	3	2
Ranunculus peltatus	–	–	1	1	1	1
Lobelia + Isoëtes	–	–	1	1	1	1
Lemnid:						
Lemna minor	1	1	–	–	1	1
Total number of dominants	87	87(5)	45	45(10)	132	132(15)
Number of lakes	23		15		38	

The changes of aquatic vegetation connected with the muskrat feeding and housing are mostly of a temporary nature, since in most cases helophyte vegetation recovers after the decline or migration of the muskrat populations. In some cases, however, the earlier dominating reedy helophytes can be completely destroyed. Normally, in eutrophic waters, these habitats are more or less rapidly recolonized by submerged and floating-leaved vegetation, and later by helophytes. In the recolonization phase earlier dominant helophytes can usually be replaced by other species. In oligotrophic waters eradication of aquatic vegetation can be more prolonged, and most of the helophyte zone can even disappear for decades.

Acknowledgements

The authors wish to express their thanks to Dr. Veijo Ilmavirta for valuable criticism on the manuscript. This study was supported financially by the Finnish National Research Council for Sciences.

References

Aario, L. 1933. Vegetation und Postglaziale Geschichte des Nurmijärvi-Sees. Ann. Bot. Soc. "Vanamo" 3(2): 1–132.

Akkermann, R. 1975. Untersuchungen zur Ökologie und Populationsdynamik des Bisams (Ondatra zibethica L.) II. Nahrung und Nahrungsaufnahme. Zschr. Angew. Zool. 62: 173–218.

Artimo, A. 1960. The dispersal and acclimatization of the muskrat, Ondatra zibethica (L.), in Finland. Papers on Game Res. 21: 1–101. Helsinki.

Björk, S. 1967. Ecological investigations of Phragmites communis. Studies in theoretic and applied limnology. Folia Limnol. Scand., 14, 248 pp.

Brander, T. 1949. Om bisamråttan i Finland ur naturskyddssynpunkt. Finlands Natur 8: 12–23.

Brander, T. 1951. Tre studier över bisamråttan (Ondatra zibethica L.). I. Bisamråttan som växtspridare (Ondatrokori). Acta Soc. Fauna Flora Fennica 67(3): 1–15.

Danell, K. 1977. Short-term plant successions following the colonization of a northern Swedish lake by the muskrat, Ondatra zibethica. J. Appl. Ecol. 14: 933–947.

Danell, K. 1978. Intra- and interannual changes in habitat selection by the muskrat. J. Wildl. Manag. 42: 540–549.

Danell, K. 1979. Reduction of aquatic vegetation following the colonization of a Northern Swedish lake by the muskrat, Ondatra zibethica. Oecologia 38: 101–106.

Ehrendorfer, F. (ed.) 1973. Liste der Gefässpflanzen Mitteleuropas. Gustav Fischer Verlag. Stuttgart. 318 pp. 2. ed.

Ilmavirta, V. 1980. Phytoplankton in 35 Finnish brown-water lakes of different trophic status. This volume, pp. 121–130.

Koponen, T., Isoviita, P. & Lammes, T. 1977. The bryophytes of Finland: an annotated checklist. Flora Fennica 6: 1–77.

Linkola, K. 1933. Regionale Artenstatistik des Süsswasserflora Finnlands. – Ann. Bot. Soc. "Vanamo" 3(5): 3–13.

Marcström, V. 1964. The muskrat Ondatra zibethica (L.) in northern Sweden. Viltrevy 2: 329–407.

Meriläinen, J. & Toivonen, H. 1979. Lake Keskimmäinen, dynamics of vegetation in a small shallow lake. Ann. Bot. Fennici 16: 123–139.

Pelikan, J., Svoboda, J. & Kvet, J. 1971. Relationship between the population of muskrats (Ondatra zibethica) and the primary production of cattail (Typha latifolia). Hydrobiologia 12: 177–180.

Rintanen, T. 1976. Lake studies in eastern Finnish Lapland. I. Aquatic flora: Phanerograms and Charales. – Ann. Bot. Fennici 13: 137–148.

Toivonen, H. & Ranta, P. 1976. Tampereen Iidesjärven vesikasvistosta ja sen muutoksista. (English Sum.: Changes in the aquatic vegetation and flora of the polluted Lake Iidesjärvi, SW Finland). – Luonnon Tutkija 80: 129–138.

EFFECTS OF THE GROWTH OF *ELODEA CANADENSIS* MICHX. IN A SHALLOW LAKE (LAKE TÄMNAREN, SWEDEN)

M. WALLSTEN

Abstract

Elodea canadensis Michx. was studied in the laboratory in order to see how it takes up phosphorus from the sediment and how it pumps it through the stem for later translocation to the surrounding water. ^{32}P was injected into the sediment.

Two stations, one with dense Elodea vegetation and the other with solitary plants, were studied in the shallow Lake Tämnaren in order to follow the influence of Elodea growth on nutrients in the water and sediment. Total phosphorus and total nitrogen were analysed in the water, sediment and in different parts of the plant. All material was taken from the same sampling point. The total phosphorus content in the water was higher most of the year at the station with dense Elodea compared to the station with sparse vegetation.

Introduction

Elodea canadensis is not a genuine European aquatic plant. The first time *Elodea* was observed in Europe was in 1836 in Ireland. The plant soon spread across Europe. It was first observed in Sweden in 1870 (Selander, 1957).

If a lake provides favourable conditions for Elodea growth then it will suppress other submerged and floating-leaved vegetation as it does not have any natural competitors. It seems that initially the plant has an immense production, lasting several years, after which it is more moderate in its production.

There are several studies of submerged vegetation dealing with the efficiency of nutrient uptake by different parts of plants. Sutcliffe (1962) mentioned that the leaves are most efficient in absorbing nutrients from the lake-water. Recent investigations (McRoy & Barsdate, 1970; Chapman *et al.*, 1974; Mayes, McIntosh & Andersson, 1977) showed that the roots are the most important in nutrient uptake. Carignan & Kalff (1979) found that all the species of submerged macrophytes tested in a lake obtained their P from the sediments and not from the lake-water (De Marte & Hartman, 1974) but the plants are more dependent on what kind of sediment they grow in (Olsen, 1954). Elodea prefers soft deep silts (Chapman *et al.*, 1974) and when small shoots come into contact with such substrate they grow very rapidly.

Some investigations show that submerged vegetation has the ability to take up nutrients through the roots, pump it through the stem, and translocate it to the surrounding water (McRoy, Barsdate & Nebert, 1972; De Marte & Hartman, 1974). Generally, the submerged vegetation is important for the nutrient circulation in lakes (Wetzel, 1975; Marshall & Westlake, 1978).

This investigation deals with the question of how phosphorus is distributed in the plant and to what extent phosphorus is translocated to the surrounding water. In addition, a study was made on the effect of a heavy growth of Elodea on the phosphorus content of lake-water.

The lake

Tämnaren is 38 km^2 with a mean depth of 1 m and a maximum depth of 1.7 m. The mean conductivity of the lake-water is 260 μS/cm (Wallsten, 1977). The lake is eutrophic having mean values of total-N of 1.5 mg/l and total-P of 100 μg/l.

The sediment in L. Tämnaren consists mostly of soft silt several meters deep. In areas where Elodea has grown for some years a gyttja sediment is found. It has a high content of organic material and a dark brown colour.

Elodea canadensis was first observed in L. Sörsjön, the southern part of Tämnaren, in about 1920. Sparse vegetation was observed in 1969 in the northern part of the lake. In 1974 some areas were completely covered by Elodea.

Flowering *Elodea canadensis* is not common in larger Swedish lakes, perhaps because the summer is not warm enough. However, in L. Tämnaren Elodea flowers were found in many places during the summer of 1975.

Material and methods

The first part of the investigation was carried out in the laboratory. ^{32}P was injected into the sediment in glass-vessels and then Elodea was planted into the sediment. Water-samples were taken at different times to measure the radioactivity in the water and the experiment ended with the harvesting of the Elodea plants.

Laboratory studies

Sediment and plant material were taken from the same place in L. Tämnaren. A fifteen cm thick layer of sediment was placed in a fifty cm high glass cylinder, with a bottom area of 380 cm^2. 1 ml of carrier-free ^{32}P solution, radioactivity of 2 m Ci, was injected into and carefully mixed with the sediment. A fine hole was made in the bottom of a polythene bag. The Elodea roots were put through the hole. A thick layer of lanolin was placed around the hole and roots, both inside and outside the bag. The roots were placed in the sediment and the bag was filled with lake-water, Fig. 1. The cylinder was placed in a room with a constant temperature of +15°C.

The method was tested by using a glass-rod instead of a plant. The investigation with ^{32}P and Elodea was performed in three cylinders with one plant in each.

The radioactivity in the water was measured eight hours after the experiment started and nine, ten and eleven days later. On each occasion three water-samples of ten ml. were taken close to the Elodea plant and also close to the cylinder wall. The radioactivity was measured by liquid scintillation using Intertechnique SL 30. The Cherenkov light emitted from an aqueous solution containing ^{32}P was detected.

The Elodea plants were harvested the same day the last water-samples were taken. They were carefully rinsed with tap-water to remove adherent radioactive water. The plants were dried on a blotting paper for two hours and then cut into pieces as the lower part of Fig. 1 shows. The pieces were weighed and put in scintillator-bottles with teflon packing in the cover. The organic material was treated with 1.5 ml HClO$_4$ and 3 ml H$_2$O$_2$ for fifteen hours at 90°C (modified after Mahin & Lofberg, 1970). The solution was measured in a liquid scintillation photometer. The material from one Elodea plant was destroyed when it was prepared for measuring.

The radioactivity in the water is given as a mean value of the three cylinders. CPM/ml are recalculated to DPM/ml H$_2$O and corrected for background and decay. The plant material was calculated in the same way, but the values given for the two plants are not mean values.

Field studies

The two sample-points in the lake were selected so that Stn A had solitary small Elodea plants and Stn B was filled with Elodea. The water depth of both stations was 0.5 m. Samples of water, sediment and *Elodea canadensis* were taken at these stations once a month from July 1974 to June 1975, except December and January.

The sediment-cores were immediately divided into 5 cm thick layers from 0 to 15 cm. All sediment strata were analysed for total-P, total-N and organic content. Density and Fe content were analysed only in 1974.

Twenty fresh Elodea plants were harvested at both sites. The plant-material was rinsed in tap water and, finally, in distilled water. Roots, dormant apices and small lateral shoots were separated from ten plants. This material was dried for twenty-four hours at 105°C. The material was

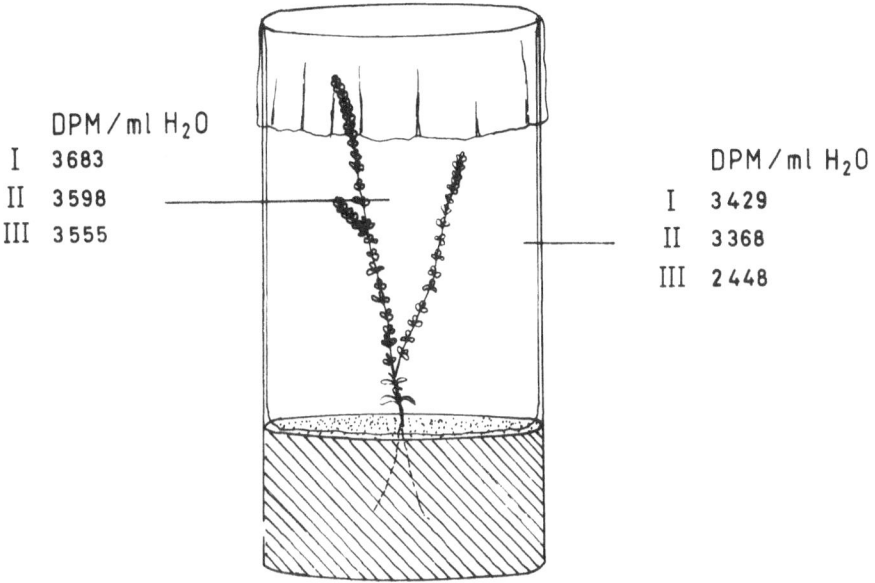

DPM/ml H₂O

	DPM/ml H₂O			DPM/ml H₂O
I	3683		I	3429
II	3598		II	3368
III	3555		III	2448

Fig. 1*a.* Vessel containing an *Elodea canadensis* plant and mean values of ^{32}P, expressed as DMP. ml H₂O^{-1}, measured near the plant and the wall of the vessel.

DPM/mg wet substance

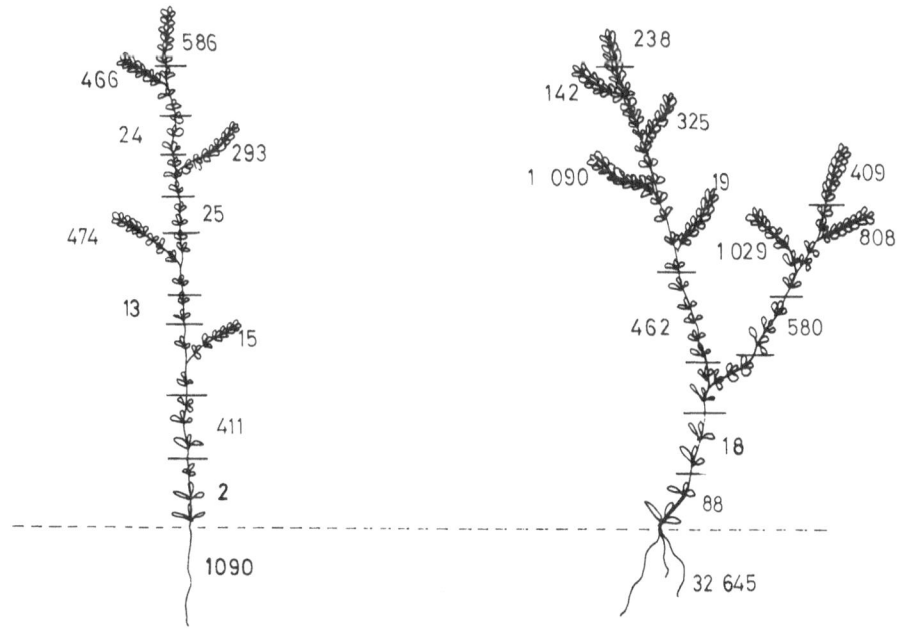

Fig. 1*b.* ^{32}P expressed as DPM/mg wet substance. in different parts of two specimens of Elodea.

ground and dried again for one hour before analysis. The plant-material was analysed for total-P and total-N. All analyses follow the methods given by Ahlgren and Ahlgren (1974). The results of sediment and plants are mean values of three analyses taken from the same sample.

Results

Laboratory studies

Young Elodea plants with many branches were used in the experiments. Young plants are more effective in taking up nutrients (Boyd, 1970). They need nutrients for their growth and probably do not translocate nutrients to the surrounding water to the same degree as older plants.

No activity due to ^{32}P was detected eight hours after the start. Water-samples taken nine days after the start of the experiment showed that the ^{32}P had increased, see Fig. 1a. The mean radioactivity counts from days nine (I), ten (II), and eleven (III) showed higher values of ^{32}P in the water close to the plants than close to the wall of the vessel. At both sites the ^{32}P values decreased; from 3683 DPM per ml H_2O to 3555 near the plant and from 3429 to 2448 DPM per ml H_2O near the wall.

The plant material was harvested eleven days after the start. Figure 1b gives the values of ^{32}P in DPM per mg wet substance from different parts of the plants. The results show higher values in the branches compared to the stem parts. The roots had the highest ^{32}P content. De Marte & Hartman (1974) did not find the same distribution of ^{32}P in the plants in their experiments with *Myriophyllum exalbescens* and *Elodea canadensis*. They found more radioactivity in the parts near the roots and at the nodes rather than in the tips.

The mean radioactivity in the water on the last day of the experiment was 3002 DPM per ml water, which gives $24 \cdot 10^6$ DMP for the 8 l water in the vessel. All the Elodea material included $66 \cdot 10^4$ DMP. The ratio between ^{32}P in the water and Elodea material was 36:1 and shows that there was a large translocation of ^{32}P from the plant to the water.

Field studies

Phosphorus content in Elodea plants in the lake varied between 0.2% and 0.7% of d.w. (dry weight).

The values in the different parts of the plant varied during the year and also between the two stations, Fig. 2. The differences between maximum and minimum values are greater at Stn B, with dense Elodea, than at Stn A. At Stn B maximum values in all three parts of the plant are found in May.

Total-P in the sediment varied between 0.06–0.1% of d.w. at both stations. There was little difference between the two sampling points. In spring total-P in the water was about twice as high at Stn B as at Stn A.

Total-N in Elodea was higher in spring than in autumn, Fig. 3. This is in agreement with results found by Best (1977). In stem, shoot and root the maximum nitrogen content was found in May. The maximum nitrogen content in Elodea was found in the shoot with 5.2 mg N per 100 mg d.w. The mean value for the whole plant is 2.2 mg per 100 mg plant material d.w. Bytniewska (1977) measured 4.76 mg N per 100 mg Elodea material as a mean value.

From August to November the nitrogen content in both stem and shoot was higher at Stn B (with dense vegetation) than at Stn A. The nitrogen content in the roots, on the contrary, was highest at Stn A.

The sediment at Stn B contained more organic material (13%) than at Stn A (8%). This reflected in the nitrogen content of the sediment, see Fig. 3. There was a maximum in August (1.5%) followed by a decrease, and in October there was a second maximum (1.2%).

The total nitrogen in the water was high in late winter at both stations, but it decreased in late spring.

Discussion

The laboratory study demonstrated an uptake of ^{32}P by *Elodea canadensis* roots and a translocation to the surrounding water. ^{32}P in the water decreased with time. An epiphytic algal production may have started taking up ^{32}P, or absorption may have occurred onto the plastic wall. It is easier to prove the translocation of phosphorus from sediment to water in a laboratory investigation than to demonstrate it in the field. Submerged plant species have been used to

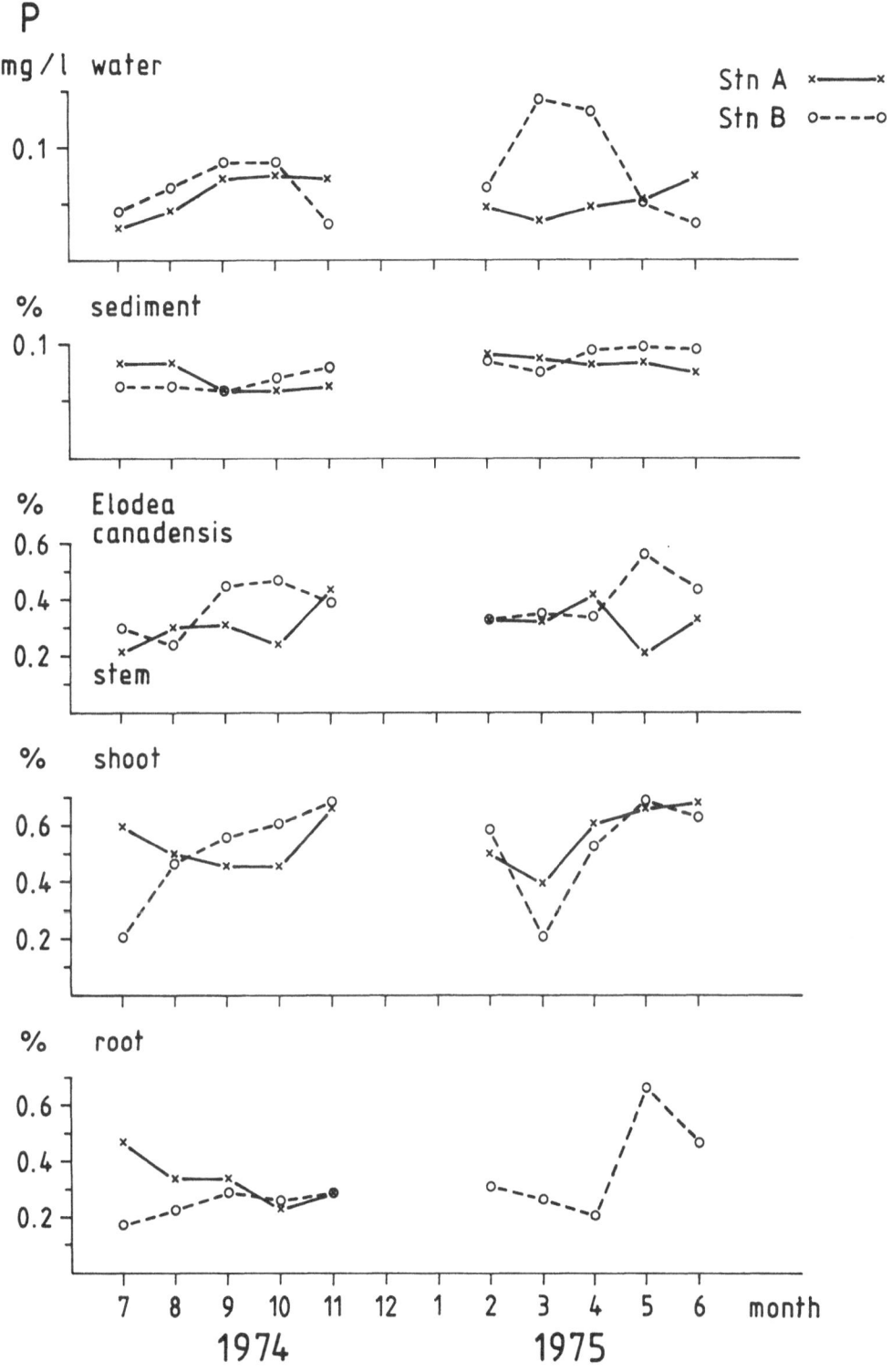

Fig. 2. Total-P at Stns A and B during 1974 and 1975. In water (mg/l); stem, shoot and root of *Elodea canadensis* and sediment (% of dry weight).

143

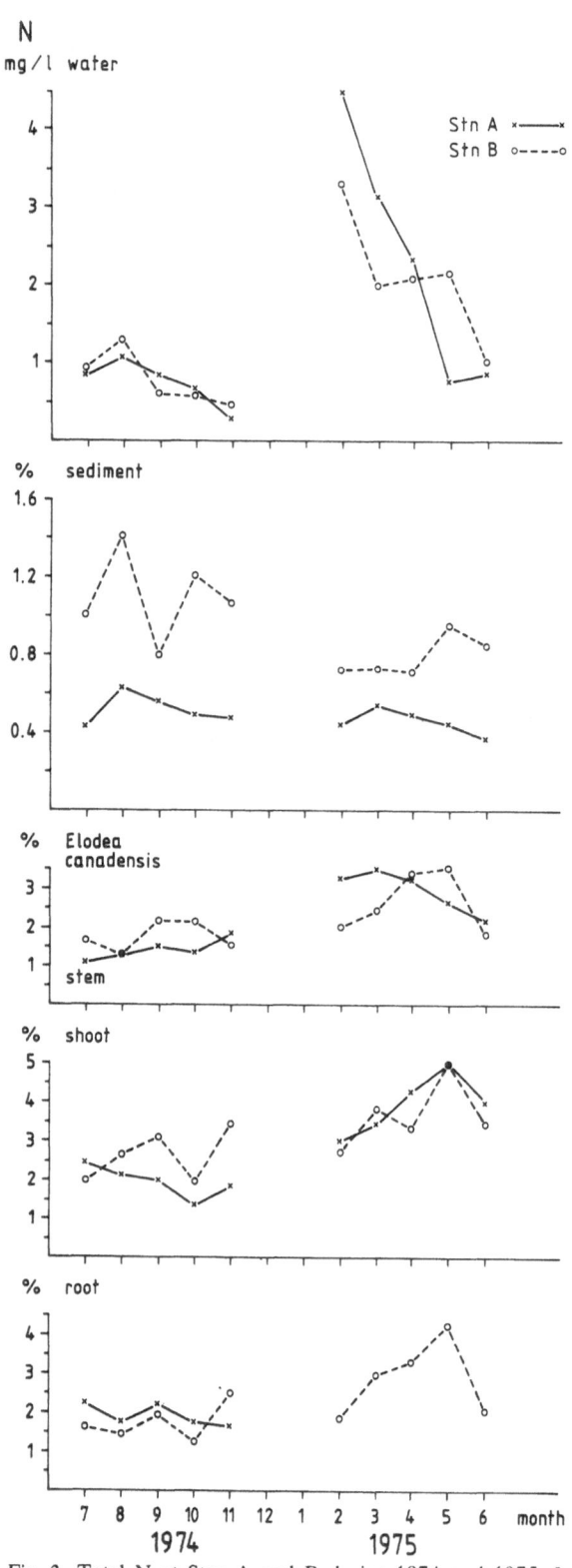

Fig. 3. Total N at Stns A and B during 1974 and 1975. In water (mg/l); stem, shoot and root of *Elodea canadensis* and sediment (% of dry weight).

demonstrate (Best & Mantai, 1978; Carpenter & Adams, 1979) nutrient uptake in many laboratory studies, but only a few field studies have been made (Peltier & Welch, 1970; Ryan, Riemer & Toth, 1972).

The nutrient circulation by plants may have a great importance (Mayes & McIntosh, 1977) in shallow lakes with a canopy of submerged vegetation. In lakes with Elodea vegetation, parts of the plants fall down to the sediment and decompose even during the growing season. The entire Elodea vegetation does not collapse in autumn. The plants looked rather unaffected by the cold water in late autumn or under ice in winter. They are a little more brownish during winter than in the summer-time. In spring, with heavy wind and wave action, a large proportion of the vegetation breaks up and dies. The bottom and shore then become covered with decaying vegetation. The nutrients in these plants are released into the water, or are taken up directly by the small green shoots, which give rise to new plants.

The canopy of submerged vegetation affects the light conditions in the lower strata. Light is important for nutrient uptake (Ingold, 1936). Haag & Gorham (1977) found in their investigation with Elodea that a low rate of net assimilation led to increasing stem buoyancy and a corresponding increase in light at the top of the canopy. As a result, the net productivity increased rapidly.

Carignan & Kalff (1979) found that the most important nutrient uptake is by the roots. The laboratory experiments showed that the largest nutrient content was in the roots but this was not seen in the lake, where the various plant parts had nearly the same nutrient content. During the growing season there may be a decrease of the nutrient content in the sediment caused by heavy production.

The nutrients translocated to the water may be taken up by young growing plants or by algae. The algal production in Lake Tämnaren is low in the area with dense Elodea vegetation. In the canopy of Elodea there may develop some kind of antipathy between plants and algae, or light intensity may be too low (Bell & Eaton, 1976). In a dense Elodea vegetation nutrients are growth limiting (Salisbury, 1963 and Sculthorpe, 1967). Adams *et al.* (1973) show that Elodea production increases if there is an input of nutrients but continued fertilization will promote algal production and submerged vegetation decreases, or disappears (Mulligan *et al.*, 1976). Dense growths of phytoplankton prevent the development of rooted species simply by light limitation.

The investigation in Lake Tämnaren shows that the total-P content in the lake water is higher at the station with dense Elodea for most of the year compared to the station with sparse vegetation, thus supporting the hypothesis that Elodea may act as a pump of nutrients between sediment and water. The high phosphorus content in the water in March and April, when the lake was still covered with ice, may be due to release from the sediment. There was no anoxia in the water, but anoxic conditions may develop close to the bottom because of the decomposition of organic material in the sediment. In February and March the phosphorus content in the sediment at Stn B, with dense vegetation, is lower than at Stn A. The phosphorus content in the sediment increased from September at Stn B. The reason is that the fresh plant material, broken up by the greater wave action at this time, stays on top of the sediment and is not immediately decomposed.

The nitrogen content of the sediment shows a considerable difference between the two sites. Nitrogen is mainly bound to organic material and the high content of organic material at Stn B gives this difference.

All three plant parts have a greater nutrient content in spring than in autumn. Both nitrogen and phosphorus in the plant show maximum values in May. Haag & Gorham (1977) found that the plants had a large nutrient uptake in spring when the growth rate was slow. The uptake depends on the concentration of available nutrients in the sediment. In a shallow lake, like Tämnaren, the nutrient circulation produced by submerged vegetation may be high. If there is a very rich production of aquatic plants, the nutrient circulation will include the sediment, the plants and the surrounding water. Also the water will be clear. Areas with less rich macrophyte production may have a larger phytoplankton production and the nutrient circulation will occur to a greater extent in the free water.

References

Adams, F. S., Cole, H. & Massie, L. B., 1973. Element constitution of selected aquatic vascular plants from Pennsylvania; submerged and floating-leaved species and rooted emergent species. Environ. Pollut. 5: 117–147.

Ahlgren, I. & Ahlgren, G. 1974. Vattenkemiska analysmetoder. Limnologiska institutionen, Uppsala.

Bell, D. & Eaton, J. W. 1976. The growth of algal epiphytes on Elodea canadensis Michx. British phycological journal, Vol II, 2: 191.

Best, E, 1977. Seasonal changes in mineral and organic components of Ceratophyllum demersum and Elodea canadensis. Aquat. Bot. 3: 337–348.

Best, M. D. & Mantai, K. E. 1978. Growth of Myriophyllum, sediment or lake water as the source of nitrogen and phosphorus. Ecology, 59: 1075–1080.

Boyd, C. E. 1970. Vascular aquatic plants for mineral nutrient removal from polluted waters. Econ. Bot. 24: 95–103.

Bytniewska, K. 1977. Nitrogen and protein contents in some aquatic plants species. Acta societatis botanicorum poloniae, Vol 46, 2: 165–172.

Carignan, R. & Kalff, J. 1979. Quantification of the sediment phosphorus available to aquatic macrophytes. J. Fish. Res. Board Can. 36: 1002–1005.

Carpenter, S. R. & Adams, M. S. 1979. Effects of nutrients and temperature on decomposition of Myriophyllum spicatum L. in a hard-water eutrophic lake. Limnol. Oceanogr. 24: 520–528.

Chapman, V. J., Brown, J. M. A., Hill, C. F. & Carr, J. L. 1974. Biology of excessive weed growth in the hydroelectric lakes of the Waikato River, New Zealand. Hydrobiologia, Vol 44, 4: 349–363.

De Marte, J. A. & Hartman, R. T. 1974. Studies on absorption of ^{32}P, ^{59}Fe, and ^{45}Ca by water-milfoil (Myriophyllum exalbescens Fernald). Ecology 55. 188–194.

Haag, R. W. & Gorham, P. R. 1977. Effects of thermal effluent on standing crop and net production of Elodea canadensis and other submerged macrophytes in Lake Wabamun, Alberta. J. appl. Ecol. 14: 835–851.

Ingold, C. T. 1936. The effect of light on the absorption of salts by Elodea canadensis. New phytologist, Vol 35, 2: 132–141.

Mahin, D. T. & Lofberg, R. T., 1970. Determinations of several isotopes in tissue by wet oxidation. Curr. Status Liquid Scintillation Counting. Ed. Bransome E. O. Grune and Stratton, New York.

Mayes, R. A., McIntosh, A. W. & Anderson, V. L. 1977. Uptake of cadmium and lead by a rooted aquatic macrophyte (Elodea canadensis). Ecology 58: 1176–1180.

McRoy, C. P. & Barsdate, R. J. 1970. Phosphate absorption in eelgrass. Limnol. Oceanogr. 15: 6–13.

McRoy, C. P., Barsdate, R. J. & Nebert, M. 1972. Phosphorus cycling in an eelgrass (Zostera marina L.) ecosystem. Limnol. Oceanogr. 17: 58–67.

Marshall, E. J. P. & Westlake, D. F. 1978. Recent studies on the role of aquatic macrophytes in their ecosystem. Proceeding EWRS 5th Symp. on Aquatic Weeds, pp 43–51.

Mulligan, H. F., Baranowski, A. & Johnson, R. 1976. Nitrogen and phosphorus fertilization of aquatic vascular plants and algae in replicated ponds. I. Initial response to fertilization. Hydrobiologia, Vol 48, 2: 109–116.

Olsen, C. 1954. Hvilke betingelser må vaere opfyldte, for at Helodea canadensis kan opnå den optimale udvikling, der er årsag til dens massevise optraeden i naturen? Botanisk Tidskrift 51: 263–273.

Peltier, W. H. & Welch, E. B. 1970. Factors affecting growth of rooted aquatic plants in a reservoir. Weed Sci. 18: 7–9.

Ryan, J. B., Riemer, D. N. & Toth, S. J. 1972. Effects of fertilization on aquatic plants, water, and bottom sediments. Weed Sci. 20: 482–486.

Salisbury, E. J. 1961. Weeds and aliens. Collins, London.

Sculthorpe, C. D. 1967. The biology of aquatic vascular plants. Edward Arnold, London.

Selander, S. 1957. Det levande landskapet i Sverige. Bonniers Förlag, Stockholm.

Sutcliffe, J. F. 1962. Mineral salt absorption in plants. Pure and applied biology: Plant Physiology, Vol 1. Pergamon Press, London, pp 1–194.

Wallsten, M. 1977. Tämnarens vattenkemi. Vatten 4: 419–427.

Wetzel, R. G. 1975. Limnology. Saunders Company, Toronto.

LAKE LA CALDERA, SIERRA NEVADA, SPAIN

BACKGROUND DATA

Latitude: 37°3'N
Longitude: 3°18'W
Altitude: 3.050 m
1(km) 0.20
b(km) 0.15
L(km)
L_D
Name of the main tributary –
Average inflow m^3/sec. –
Average outflow m^3/sec.–
Theoretical retention time

origin: Glaciar
$z(m) \simeq 12$
$\bar{z}(m) \simeq 4.7$
$\bar{z}/z = 0.4$
$V(km^3)$ 107.6×10^{-6}
$A(km^2)2.3$
$A'(km^2)$
$A':A$

Geological characteristics

Metamorphic materials made out of sediments and volcanic rocks.
Main materials of the lake's basin are schists, marbles and amphibolites.

Climatic conditions

Average monthly temperature 6°C
Average precipitation/year 40.8 lm^{-2}
Average sunshine duration
Main wind direction(s)
Evaporation per year 1m^3/m^2

Ice cover (days) $\simeq 200$
Average radiation/year
% of calm days

Cultural geography and demography:

Land usage of catchment area (%) –

Industrial
Agricultural
Meadows
Forest

High mountain zone
Usage of lake water including –
recreation activities

unused ×
number of residents   inhabitants/km^2
Water temperature: min 0°C   Max 12.5°C
Secchi depth: min   max total
Euphotic zone: min   max total
O$_2$ concentration: min 5.9 mg · l^{-1} max 9 mg · l^{-1}
pH 7   Conductivity (μS) –   Alkalinity (mval) 13 to 26 meq · l^{-1}
Average P-conc. –   Average N-conc. 8 μg-at · NO$_3$ – N · l^{-1}
Conditions of sediment:

Dom. phytoplankton species:
   Chromulina mikroplancton, Ch. rosannoffi, Ch. parvula, Chrysococcus minutus, Cyanarcus hamiformis,
   Rhodomonas minuta var. nannoplanctia

Dom. zooplankton species:
   Mixodiaptomus laciniatus, Diaptomus cyanaeus

Dom. macrophytes:

Dom. benthic organisms
   Navicula, Cymbella and Cocconeis species

Fishes:
   no

LAKE TÄMNAREN, SWEDEN

BACKGROUND DATA

Latitude: 60°10′N
Longitude: 17°20′
Altitude: 35 m.a.s.
1(km) 9.2
b(km) 5.2
L(km) 50
L_D
Name of the main tributary Harboån, Åbyån
Average inflow m³/sec. 3.7
Average outflow m³/sec. 5.1
Theoretical retention time 2 months

origin:
z(m) 1.7
$\bar{z}$(m) 1.0
$\bar{z}/z$ 0.6
V(km³) 0.038
A(km²) 719
A′(km²) 38
A′ : A   0.05

Geological characteistics

A lake against a fault scarp. The fault is heavily eroded. The whole region is flat country.

Climatic conditions

Average yearly temperature + 6°C
Average precipitation/year 600 mm
Average sunshine duration 1895 h
Main wind direction(s) SW (19.7%)
Evaporation per year 375 mm

Ice cover (days) 155
Average radiation/year 80000 cal
% of calm days 11.6

Cultural geography and demography:
Land usage of catchment area (%)

Industrial –
Agricultural ⎫
            ⎬ 149 km² (21)
Meadows   ⎭

Usage of lake water including
recreation activities
    Water reservoir
    Fishing

Forest   560 km² (72)
unused
number of residents   inhabitants/km² 10
Water temperature: min 0.9°C   max 18.5°C
Secchi depths: min 0.5 m   max bottom
Euphotic zone: min 0.6 m   max bottom
O₂ concentration: min 8.8 mg/l   max 20.2 mg/l
                                 96%   152%

pH 8   Conductivity (μS) 260   Alkalinity (mval) 1.5
Average P-conc. 97 μg/l (total-P) Average N-conc. 1.5 mg/l (total-N)
Conditions of sediment:
  Clay – mud. The top-5 cm has a organic content of about 10%.

Dom. phytoplankton species:
  Tabellaria

Dom. zooplankton species:
  Polyarthra vulgaris, Filinia longiseta, Bosmina longirostris

Dom. macrophytes:
  Phragmites australis

Dom. bentihic organisms:

Fishes:
  Perca fluviatilis, Abramis brama

4.3 Bacteriological Aspects

DIURNAL VARIATIONS IN THE UPTAKE OF GLUCOSE BY ATTACHED AND FREE-LIVING MICROHETEROTROPHS IN LAKE WATER (LAKE MOSSØ, DENMARK)

B. RIEMANN

Abstract

Diurnal variations in glucose uptake in different size fractions of freshwater microheterotrophs were measured during two twenty-four hour surveys in the eutrophic lake Mossø, Denmark. Sampling at three-hour intervals revealed 3–4 fold irregular variations in glucose uptake. The percentage uptake (percentage of total uptake) in one size fraction was relatively constant throughout the surveys. Single sampling is discussed in relation to diurnal variations in glucose uptake and the distribution of attached and free-living microheterotrophs in eutrophic lake water.

1. Introduction

Heterotrophic microorganisms, mainly bacteria and fungi, are important factors in the decomposition of organic matter and regeneration of nutrients in lake water.

To understand the dynamic importance of heterotrophic organisms, it is often a prerequisite to estimate their biomass and/or production. Although many methodological improvements have been made in the determination of biomass of aquatic microorganisms (e.g. Azam & Hodson, 1977; Riemann, 1978b), such measurements are still too inaccurate to allow reliable calculations of production to be made.

Despite the fact that glucose usually only constitutes a small fraction (<1–11%) of total dissolved organic carbon (Allen, 1973; Berman et al., 1979), it may still be recognized as an important compound in the rapid cycling of low molecular weight compounds in lake water. Further, when corrections are made for respiration (Hobbie & Crawford, 1969; Cavari & Hadas, 1979), data of glucose net assimilation may be related to the flux of organic carbon (Berman et al., 1979).

This study is part of a project designed to evaluate the role of microheterotrophs in eutrophic lake water. Short term in situ uptake experiments using glucose were made in attempts to determine the diurnal variations in the activity of attached and free-living microheterotrophs in a shallow eutrophic lake.

2. Materials and Methods

Water samples were taken with a Ruttner water sampler (1.8 liter) at a depth of 1 m in the western part of Lake Mossø (Riemann, 1977).

Twenty-four-hour surveys were carried out from the 5th to the 6th of October, 1978 and the 4th to the 5th of September, 1979. Glucose uptake, total particulate ATP and chlorophyll a were determined every 3 hours.

Glucose uptake experiments were carried out in situ. Additions of 0.5 μCi ^{3}H-D-glucose (22.5 Ci/mmol, Amersham, England), corresponding to about 4 ng glucose, were made to 100 ml aliquots of lake water in clear Jena bottles. For details of method, see Riemann (1979a).

ATP was determined by the method of Holm-Hansen & Karl (1978) except for the enzyme preparation, which was made according to Karl & LaRock (1975). Samples of total particulate ATP, corrected for dissolved ATP (Riemann, 1979a), were obtained from samples of $2 \times 500 \mu l$ which were added directly to the boiling Tris buffer. In the quantification of sample ATP, corrections were made for non-ATP nucleotide interference by means of proper dilutions of crude firefly enzyme and subsequent quantification of light emission by means of short term (0–6 s) integration periods (Karl, 1978; Riemann, 1979b).

Chlorophyll *a* (not corrected for phaeopigments) was extracted in absolute methanol according to Holm-Hansen & Riemann (1978). The absorption coefficient of 77.9 liter g^{-1} cm^{-1} for chlorophyll *a* in methanol was applied to calculate the concentration (Riemann, 1978a).

The use of Nuclepore filters in size fractions of natural populations of microorganisms was studied by filtering lake water containing radiolabelled microheterotrophs through filters with varying pore sizes (3.0, 1.0, 0.6, 0.4 and 0.2 μm). Each filter was washed with about 10 ml distilled water and assayed by LS-counting.

The effect of formalin on the cellular pool of radiolabelled material was tested in uptake experiments using ^{3}H-D-glucose and ^{3}H-ATP. Immediately after 30 min incubation in the laboratory at the ambient temperature, 20 ml aliquots were filtered through 0.2 μm Nuclepore filters. At the same time, formalin was added to similar 20 ml aliquots (final concentration 2%), and these samples were stored for varying periods of time at room temperature before filtration.

3. Results

In Lake Mossø, it is normally sufficient to filter 20–30 ml lake water to give reliable measurements of activity in the different size fractions (Riemann, 1978b). Sometimes, however, it is necessary to decrease or increase this volume of water due to differences in phytoplankton density. On several occasions the effect of using different volumes of water was tested in order to evaluate possible changes in the distribution of activity in the different size fractions. A representative example is shown in Fig. 1. Volumes of water between 15 and 50 ml caused no marked changes in

the distribution of glucose uptake, except for the activity on the 3.0 μm filters, where a decrease in activity was found when volumes of water were increased from 15 to 50 ml.

When measuring heterotrophic activity, up to an hour could often elapse between sampling and assay, requiring the additions of a preservative to terminate the uptake of glucose by the microheterotrophs. The effect of additions of formalin (2% final concentration) on the cellular pool of radioactive material incorporated was tested in experiments using glucose and ATP as substrates (Fig. 2). No significant decrease in activity was found after 60 min using glucose and ATP.

The diurnal variations in glucose uptake of attached ($>0.6 \mu m$) and free-living ($<0.6 \mu m >$ $0.1 \mu m$ or $0.2 \mu m$) microheterotrophs was followed during two twenty-four-hour surveys. In October 1978, the phytoplankton population was dominated by diatoms (*Melosira* and *Fragilaria*) and bluegreen algae (*Microcystis* and *Aphanizomenon*). A nearly four-fold irregular variation was observed in the glucose uptake of the free-living microheterotrophs, $<0.6 \mu m >0.1 \mu m$ (Fig. 3). The diurnal variation in the distribution of the microheterotrophs (given as activity in size fraction $<0.6 \mu m >0.1 \mu m$ as a percentage of total uptake) changed far less than did variations in the uptake of glucose.

The marked increase in chlorophyll *a* (Fig. 4) from 9 p.m. to 6 a.m. (October 6) reflects that sampling from a fixed station results in changes in the phytoplankton and probably also in the bacteria populations. As the increase in chlorophyll *a* was not paralelled by an increase in ATP, resuspension of bottom material containing dead algae is suspected.

During the second twenty-four-hour survey in September 1979, diatoms dominated completely (*Melosira*, *Asterionella* and *Fragilaria*). Again, large variations were observed in the glucose activity of the free-living microheterotrophs and the distribution of the activity was much more constant (Fig. 5).

The chlorophyll *a* concentrations were fairly constant (Fig. 6) except from a significant decrease in the early morning. The same trends were observed in the ATP content, except for a marked increase at 9 p.m.

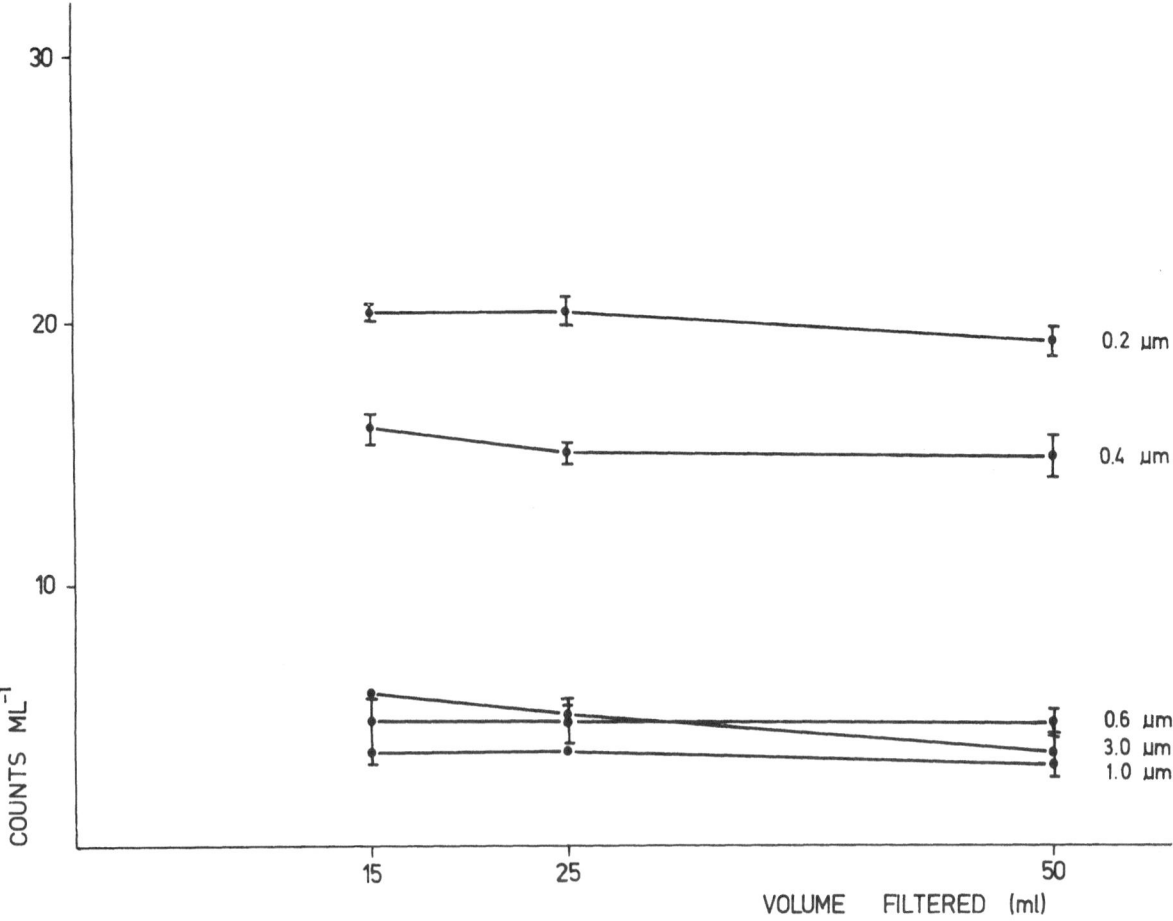

Fig. 1. Effects of filtering varying volumes of water through Nuclepore filters of various pore sizes on the distribution of microheterotrophic ^{3}H-D-glucose activity using natural populations of microorganisms from Lake Mossø. Bars indicate SE for $n = 3$.

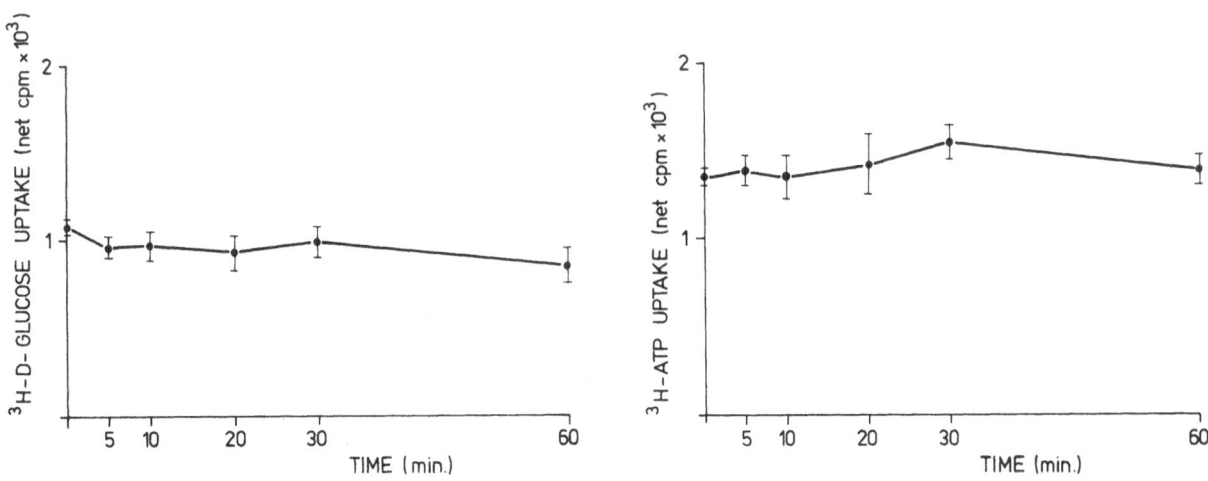

Fig. 2. Time effects of formalin (2% in final solution) on the cellular pool of incorporated ^{3}H-D-glucose and ^{3}H-ATP using natural populations of microorganisms from Lake Mossø. At time zero, aliquots without formalin were filtered through 0.2 μm Nuclepore filters and measured for activity. Formalin was added to other aliquots, and these fixed samples were filtered after varying periods of time. Bars indicate SE for $n = 3$.

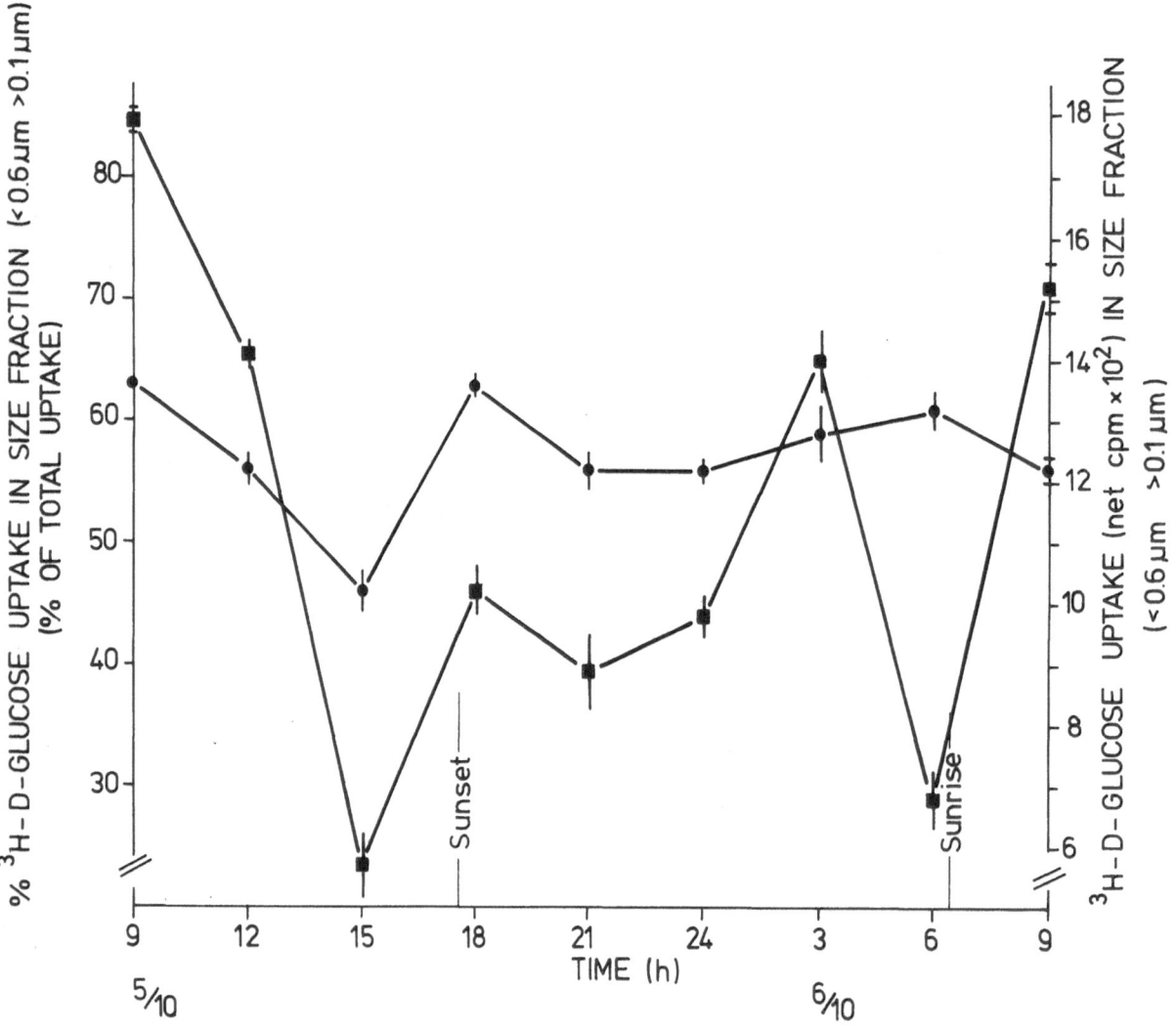

Fig. 3. Diurnal variations in ³H-D-glucose uptake in size fraction <0.6 μm >0.1 μm compared to percentage uptake in this size fraction (percentage of total uptake) in Lake Mossø October 5–6, 1978. ■—■: glucose uptake, ●—●: percentage uptake. Bars indicate total variation of two determinations.

The mean values of glucose uptake from the two twenty-four-hour surveys were calculated, and the coefficient of variation (CV) in the two size fractions were compared with CV calculated from the percentage uptake (percentage of total uptake). In the two size fractions, CV varied from 23 to 36% for glucose uptake and from 4 to 14% for percentage distribution of the microheterotrophs (Table 1).

4. Discussion

The use of filters or screens in studies of size fractionations of aquatic microorganisms requires a constant ratio between different size fractions when different volumes of water are used for filtration. Such requirements are seldom met with membrane filters, since they tend to retain particles much smaller than their rated pore size (Sheldon & Sutcliffe, 1969; Sheldon, 1972). Nuclepore filters with a pore depth of 5–10 μm have improved the size fractionation technique. As shown by the results in Fig. 1, reliable estimates of the distribution of different size fractions can be made in Lake Mossø when volumes of 15–50 ml are used for filtration.

Another necessary step in activity studies is to terminate the uptake of the added isotope by

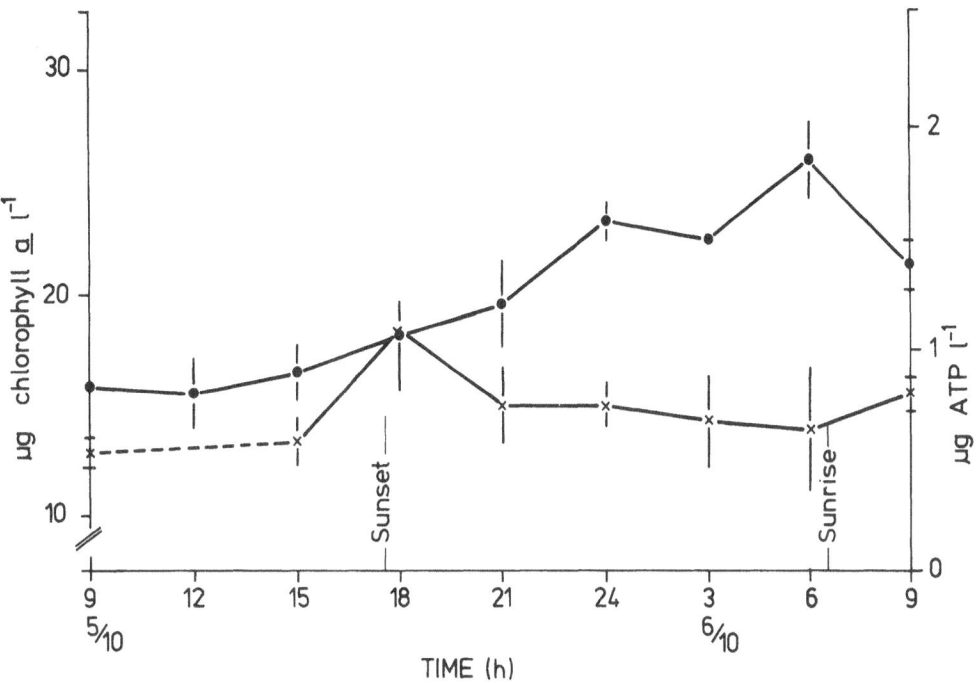

Fig. 4. Diurnal variations in chlorophyll *a* (●—●) and total particulate ATP (×—×) during a survey in Lake Mossø October 5–6, 1978. Bars indicate total variation of two determinations.

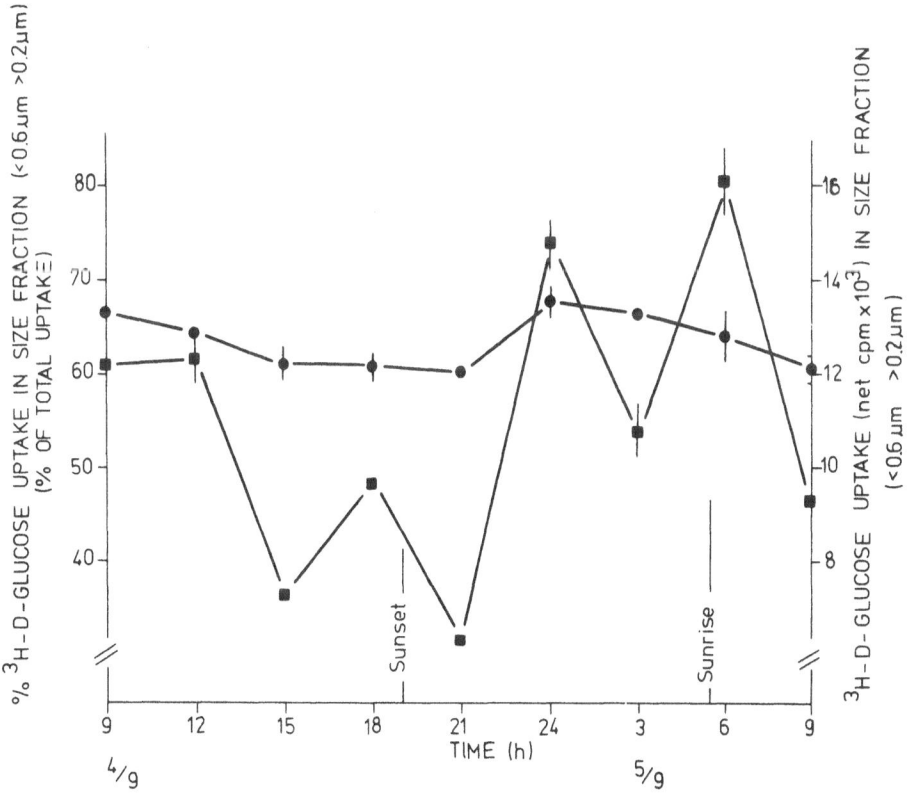

Fig. 5. Diurnal variations in ³H-D-glucose uptake in size fraction <0.6 µm >0.2 µm compared to percentage uptake in this size fraction (percentage of total uptake) in Lake Mossø September 4–5, 1979. ■—■: glucose uptake, ●—●: percentage uptake. Bars indicate total variation of two determinations.

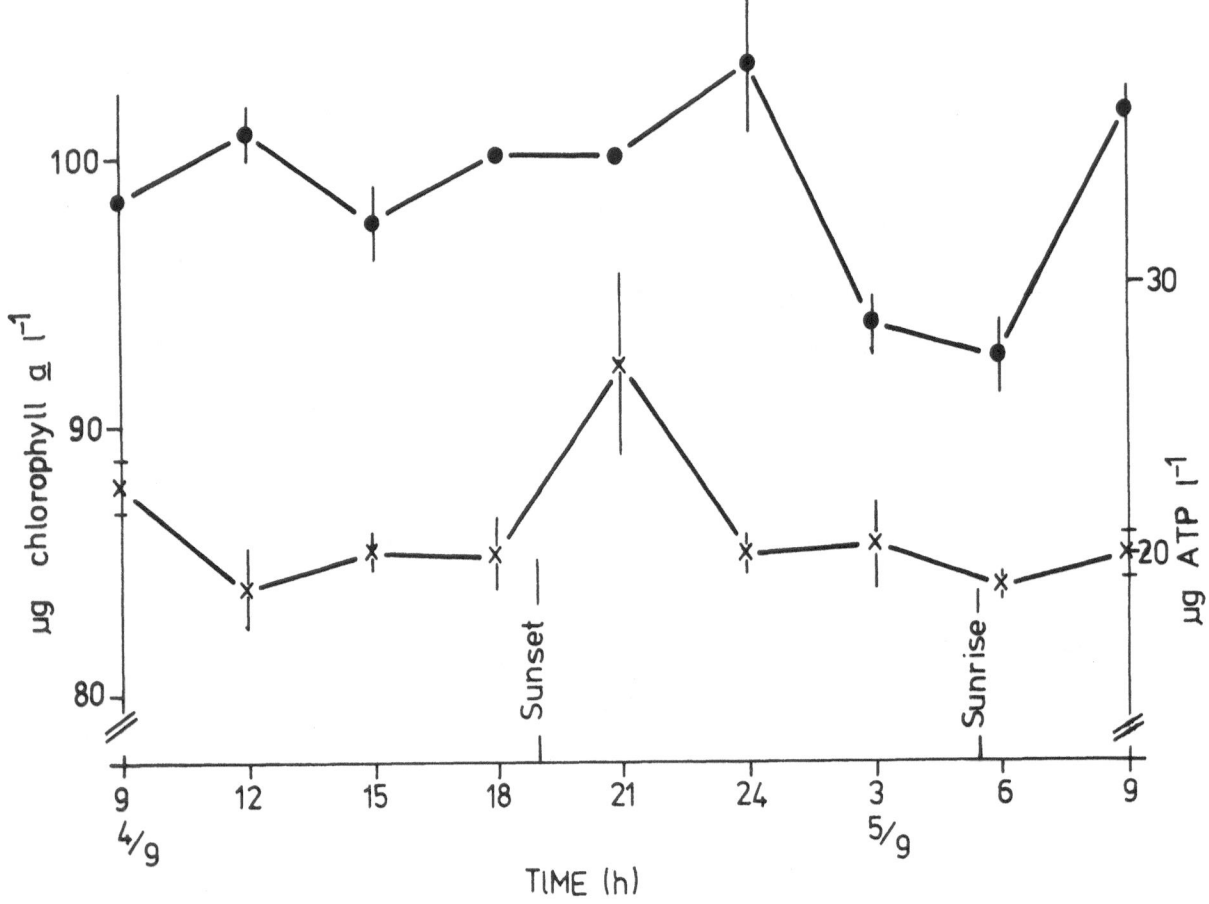

Fig. 6. Diurnal variations in chlorophyll *a* (●—●) and total particulate ATP (×—×) during a survey in Lake Mossø September 4–5, 1979. Bars indicate total variation of two determinations

killing the organisms with a preservative. Formalin may cause release of the cellular pool of radiolabelled material from phytoplankton (Il-mavirta & Jones, 1977; Silver & Davoll, 1978). The results of this study indicate that loss of cellular radiolabelled material (glucose and ATP) incorporated in freshwater microheterotrophs may be considered as insignificant, at least in samples exposed to formalin treatment for up to an hour. Similar results are reported by Allen (1969) Cavari & Hadas (1979) in studies using ^{14}C-glucose, indicating that formalin can be used in

Table 1. Values for CV (SD % of mean) calculated from the results of glucose uptake and percentage glucose uptake (percentage of total uptake) in each size fraction during the two twenty-four-hour surveys.

	^{3}H-D-glucose uptake	% of total uptake
5–6/10–78		
>0.6 μm	29	14
<0.6 μm >0.1 μm	36	9
4–5/9–79		
>0.6 μm	23	7
<0.6 μm >0.2 μm	29	4

heterotrophic uptake studies in freshwater samples.

Despite the fact that heterotrophic microorganisms utilize a varity of organic substances, glucose uptake studies can often be a proper method of elucidating natural heterotrophic activity because (1) glucose is a common sugar in many environments (Whittaker & Vallentyne, 1957; Berman *et al.*, 1979) and (2) many naturally occurring bacteria are able to utilize it (Düsing, 1973; Overbeck, 1975).

When quantitative studies of glucose uptake are required, an important assumption is that the samples should be representative for a well-defined period of time. During the two diurnal periods studied in Lake Mossø, the glucose uptake activity of both attached and free-living microheterotrophs varied irregularily by a factor of 3–4. Similar large variations were obtained in uptake studies using ^{3}H-ATP during a diatom bloom in Lake Mossø (Riemann 1979a). In Lake Kinneret, uptake rates of ^{14}C-glucose varied 10-fold during three diurnal surveys (Cavari & Hadas, 1979).

The calculated CV figures obtained from the two surveys (calculated from the nine measurements throughout each survey) were about 29% (average figure for the four values in Table 1) for the activity of both attached and free-living microheterotrophs. CV for percentage distribution was about 9% (average for the four values in Table 1). The high CV figures illustrate that the large uncertainties may be suspected if results from a single sampling are used to estimate the diurnal uptake rates. Further, it is questionable whether sampling at three-hour intervals is sufficient to reflect the actual changes in the uptake of glucose. So far, this problem has not been solved.

Considering the distribution of the microheterotrophs in different size fractions, the CV value of about 9% seems to justify measuring the percentage of attached microheterotrophs during a time period of twenty-four-hours on the basis of a single sampling. These results are surprising in view of the fact that marked variations were observed in the chlorophyll *a* concentrations (Figs. 4 and 6), indicating that different populations of microorganisms were sampled throughout the surveys. Subsequent changes in the distribution of the bacteria could be suspected.

In conclusion, further investigations are needed to evaluate short term heterotrophic activity before reliable measurements can be made on the diurnal rates of decomposition of dissolved organic matter in lake water.

Acknowledgements

I thank M. Søndergaard for reading the manuscript and D. Clayre for his linguistic suggestions. Technical assistance was given by J. Bargholz and W. Martinsen. This work was sponsored by grants no. 511-10607 and 511-10065 from The Danish Natural Science Research Council.

References

Allen, H. L. 1969. Chemo-organotrophic utilization of dissolved organic compounds by planktonic algae and bacteria in a pond. Int. Revue ges. Hydrobiol. 54: 1–33.

Allen, H. L. 1973. Dissolved organic carbon: patterns of utilization and turnover in two small lakes. Int. Revue ges. Hydrobiol. 58: 617–624.

Azam, F. & Hodson, R. E. 1977. Size distribution and activity of marine microheterotrophs. Limnol. Oceanogr. 22: 492–501.

Berman, T., Hadas, O. & Marchaim, U. 1979. Heterotrophic glucose uptake and respiration in Lake Kinneret. Hydrobiologia 62: 275–282.

Cavari, B. Z. & Hadas, O. 1979. Heterotrophic activity, glucose uptake and primary productivity in Lake Kinneret. Freshwat. Biol. 9: 329–338.

Düsing, F. 1973. Zur Stoffwechseldynamik der fliessenden Welle bakterieller Abbau gelöster organischer Komponenten. Thesis, Univ. Kiel.

Hobbie, J. E. & Crawford, C. C. 1969. Respiration correction for bacterial uptake of dissolveed organic compounds in natural waters. Limnol. Oceanogr. 14: 528–532.

Holm-Hansen, O. & Karl, D. M. 1978. Biomass and adenylate energy charge determination in microbial cell extracts and environmental samples. Methods in Enzymol. Vol LVII: 73–85.

Holm-Hansen, O. & Riemann, B. 1978. Chlorophyll α determination: Improvements in methodology. Oikos 30: 438–447.

Ilmavirta, V. & Jones, R. I. 1977. Factors affecting the ^{14}C method of measuring phytoplankton production. Ann. Bot. Fennici 14: 97–101.

Karl, D. M. 1978. Occurrence and ecological significance of guanosine triphosphate in the ocean and in microbial cells. Appl. Environ. Microbiol. 36: 349–355.

Karl, D. M. & LaRock, P. A. 1975. Adenosine triphosphate measurements in soil and marine sediments. J. Fish. Res. Bd. Can. 32: 599–607.

Overbeck, J. 1975. Ecology of aquatic organisms. I. Bacteria distribution pattern of uptake kinetic response in a stratified eutrophic lake. Verh. Intern. Verein. Limnol. 19: 2600-2615.

Riemann, B. 1977. Phosphorus budget for a non stratified Danish lake and horizontal differences in phytoplankton growth. Arch. Hydrobiol. 79: 357–381.

Riemann, B. 1978a. Absorption coefficients for chlorophylls *a* and *b* in methanol and a comment on interference of chlorophyll *b* in determination of chlorophyll *a*. Vatten 3: 187–194.

Riemann, B. 1978b. Differentiation between heterotrophic and photosynthetic plankton by size fractionation, glucose uptake, ATP and chlorophyll content. Oikos 31: 358–367.

Riemann, B. 1979a. The occurrence and ecological importance of dissolved ATP in fresh water. Freshwat. Biol. 9: 481–490.

Riemann, B. 1979b. Interference in the quantitative determination of ATP extracted from freshwater microorganisms. Proc. Symp. Anal. Appl. Biolumin. & Chemilumin. Bruxelles 1978. (Eds.) E. Schram & P. Stanley, p. 316–332.

Sheldon, R. W. 1972. Size separations of marine seston by membrane and glass fibre filters. Limnol. Oceanogr. 17: 494–498.

Sheldon, R. W. & Sutcliffe, W. H. 1969. Retention of marine particles by screens and filters. Limnol. Oceanogr. 14: 441–444.

Silver, M. W. & Davoll, P. J. 1978. Loss of ^{14}C activity after chemical fixation of phytoplankton: error source for autoradiography and other productivity measurements. Limnol. Oceanogr. 23: 263–268.

Whittaker, J. R. & Vallentyne, J. R. 1957. On the occurrence of free sugars in sediment extracts. Limnol. Oceanogr. 2: 98–110.

PHYSIOLOGICAL ADAPTATION AND GROWTH
OF PURPLE AND GREEN SULFUR BACTERIA
IN A MEROMICTIC LAKE (VILA) AS COMPARED TO A HOLOMICTIC LAKE (SISO)

R. GUERRERO, E. MONTESINOS, I. ESTEVE & C. ABELLA

Abstract

In Vila (meromictic) and Siso (holomictic) lakes, two populations of green (*Chlorobium phaeobacteroides*) and purple (*Chromatium minus*) sulfur bacteria have different ecological niches mostly determined by sulfide concentration and light climate. The total numbers, vertical distribution and physiological characteristics (specific pigment contents and carotenoid/bacteriochlorophyll ratio) of both populations, together with the main physico-chemical and morphometric parameters are compared in the two lakes. Optical properties of water, such as light penetration and turbidity, are affected by the presence of dense populations of bacteria, which strongly modify the quality and quantity of available light. Photosynthetic bacterial populations in nature adapt to decreasing available light by increasing the specific content of bacteriochlorophylls and carotenoids. This, with the ability to move vertically in the water column, makes it possible to adapt to the underwater light climate at each depth.

Introduction

It is well known from the literature that in the anaerobic layers of meromictic and holomictic lakes photosynthetic sulfur bacteria may be present. These populations frequently stain the water with pink-purple, brown or green colors (Pfennig, 1967). Sometimes, the bacteria form plates which are a consequence of the delicate balance of the conditions required for their growth. Up to the present, most field work has been done in periods of high population density, namely, during the summer stratification (Trüpper & Genovese,

1968; Takahashi & Ichimura, 1968, 1970; Kuznetsov, 1977; Bergstein *et al.*, 1979). However, the study of population dynamics is one of the few methods at the disposal of the microbial ecologist to ascertain the importance of physiological adaptations of bacteria to changing conditions in their natural environment.

The aim of this article is to compare some physiological adaptations of purple and green sulfur bacteria populations to sulfide and light gradients in two neighboring water bodies with similar climatic conditions. The water bodies studied in the present work are located in the karstic system of Banyoles Lake (Girona, Spain). They are Vila, a meromictic, crenogenic two basin lake connected by a shallow channel to the south-west part of the principal one (Banyoles), and Siso, a small, anoxic water body situated some 200 m west of Banyoles Lake. Vila and Siso are at 175 m above sea level and about 1 km apart. Photosynthetic bacterial populations in both lakes show different vertical distribution patterns and physiological adaptations (pigment content) determined by the different limnological parameters and light environments.

Material and methods

Water samples were collected biweekly by means of a Ruttner bottle. Transparency was estimated with a 30 cm Secchi disc. Light penetration was measured with a cadmium sulfide photoresistor

(Miniwatt RPY20) mounted in a white teflon diffuser. Turbidity was measured with a vertical transmission-meter, made in the course of this work, with a light path length of 15 cm and a colored filter (Kodak Wratten No. 75, maximum transmission at 480 nm) inserted inside the light detector. Temperature was measured *in situ* by a telethermometer (Miniwatt NTC resistor). Morphometric parameters were obtained from echosounding profiles and expressed following Hutchinson (1957). Conductivity was measured *in situ* by means of a Chemtrix 700 conductivity meter. Oxygen by a Chemtrix 300 oxygen meter. Chemical analysis of H_2S. SO_4^{2-}, NH_4^+, NO_3^- and PO_4^{3-} were performed by standard methods described in Golterman *et al.* (1978).

Bacterial total numbers were estimated by direct microscopy on membrane filters (Sartorius, 0.45 μm pore). Bacterial biomass is expressed in milligrams of wet weight (w.w.), calculated using the total number, the estimated electron microscopy volume and assuming 1.07 g/cm^3 as cell density. Photosynthetic pigments were extracted in 90% acetone after filtration of water samples through membrane filters, sonic disruption and centrifugation. Extracted pigments were scanned in a Pye Unicam SP1700 spectrophotometer between 350–850 nm. Bacteriochlorophylls (bchl.) *a* and *d* were calculated following Takahashi & Ichimura (1970) equations. Carotenoid concentrations are expressed in arbitrary units (A.U.) per liter:

(i) $\qquad$ A.U./l = $A_\lambda^* x v / V$

A_λ^* being the corrected absorption (see below) of each carotenoid, determined in a 1 cm cell at the indicated wavelengths; v the volume of the 90% acetone extract in milliliters; and V the volume of filtered sample in liters.

Okenone maximum absorption at 520 nm and chlorobactene + isorenieratene at 470 nm were measured, and their values were corrected for bchl. *a*, *d* and carotenoids absorptions in accordance with the following equations, in an analogous manner to the "trichromatic" method for chlorophyll (Parsons & Strickland, 1963):

(ii) corrected absorption of okenone (A_{520}^*):

$$A_{520}^* = A_{520} - 0.040A_{654} - 0.046A_{775}$$

(iii) corrected absorption of isorenieratene + chlorobactene (A_{470}^*):

$$A_{470}^* = A_{470} - 0.040A_{654}$$
$$- 0.046A_{775} - 1.000A_{520}^*$$

A_{654} being the absorption of bchl. *d* and A_{775} of bchl. *a*.

Results

Morphometry

In Fig. 1 the location and bathymetric profiles of Siso and Vila lakes are presented. Table 1 shows the main morphometric parameters of both water bodies. Each basin of Vila has similar general characteristics. These are usual features in dolines of karstic systems (Hutchinson, 1975).

The two basins of Vila have a central zone of sediment in permanent suspension which is not considered in calculating the parameters of Table 1. The sediment in suspension of the north basin (I) reaches from 9 to 33 m depth and that of the south basin (II) reaches from 9 to 18 m. The main characteristics of this sediment are discussed next in this paper.

Physico-chemical parameters and zonation

Table 2 shows the general physical and chemical trends in both lakes. Sulfate is the dominant anion, followed by high bicarbonate alkalinity. Phosphate and nitrate are almost undetectable, but ammonium is present in very high concentrations (32–35 mg/l).

The central part of each basin of Vila lake is a conic hole filled with a sediment, in permanent suspension, formed by marls, grey to blue in color, composed of particles of different sizes and with an average density of 1.35 g/cm^3. The interphase between sediment and clear water can ascend and descend depending on the water inflow into these holes, as also happens in several basins of the main lake, Banyoles. In Vila lake the interphase is around 9 m, in years of normal rainfall.

The data of Table 2 for Vila lake shows two sharp changes of conductivity and allows the demarcation of three physico-chemical zones, separated by pycnoclines: (1) the mixolimnion, an oxygenated layer which reaches from the surface to about 4–5 m depth, (2) the monimolimnion (between 4–5 to 9 m depth), an anoxic layer with high

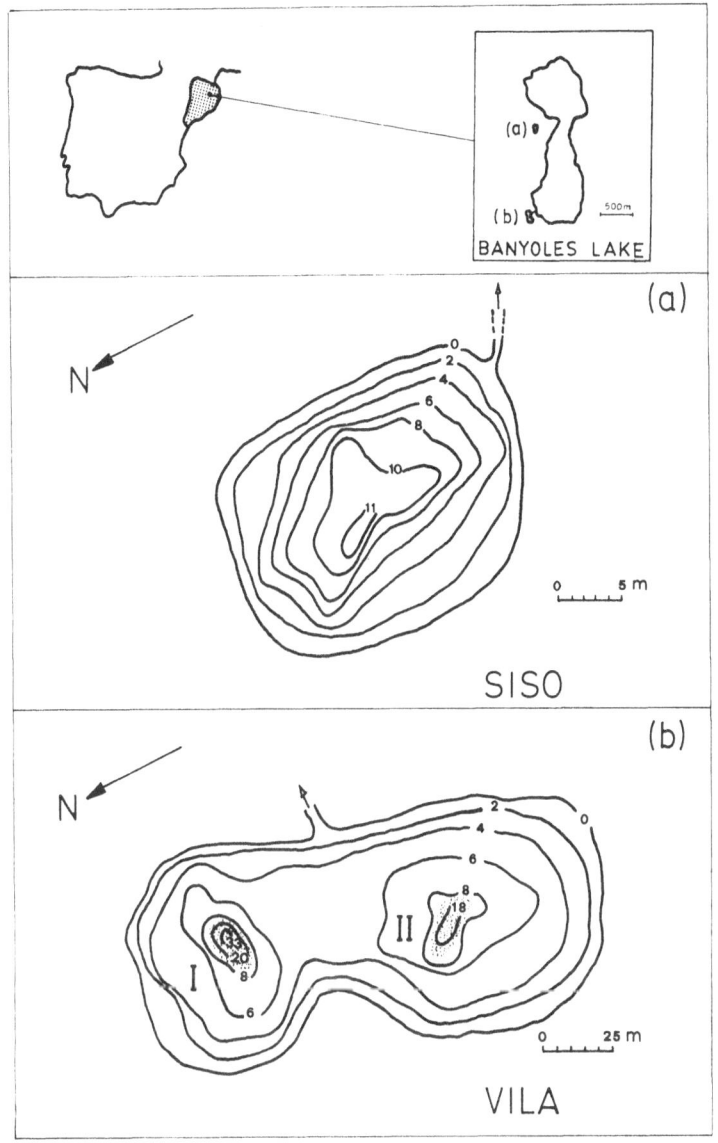

Fig. 1. Bathymetric profiles of Siso and Vila lakes. Their location in Spain and around Banyoles are indicated at the top.

sulfide concentration, which is the zone where purple and green sulfur bacteria thrive, and (3) the sediment in suspension, which occupies the central hole (about 12 m mean diameter at 9 m but becoming rapidly narrower with depth) of each basin.

On the other hand, during the year, Siso lake has two characteristic vertical distributions in the water column, as do other holomictic (monomictic) lakes of the same climate. During winter, spring and autumn the water is mixed, and the abundance of sulfide demonstrates the total anoxic

conditions up to the surface. In the summer a period of stratification is established, with an oxygenated shallow epilimnion (0 to 1 m depth) and an anoxic hypolimnion of uniform conductivity. The thermocline is sharp (3°C per meter) and has a high sulfide gradient (from 0 to 100 mg/l).

Bacterial identification and maximum total numbers

As is discussed in the accompanying paper for Siso water (Abella *et al.*, 1980), the monimolimnion of Vila has a very low diversity of photosynthetic

163

Table 1. Morphometric parameters for the water masses[*] of Siso and Vila lakes. Notation and definitions following Hutchinson (1957).

Parameter	Siso	Vila
Surface Area (A_o)	457 m^2	12 450 m^2
Total Volume (V)	2334 m^3	53 832 m^{3}[*]
Volume of mixolimnion (V_{mix})	-	47 376 m^3
Volume of monimolimnion (V_{mon})	-	6.456 m^3
Maximum depth (z_m)	11 m	9 m[*]
Length (1)	26 m	170 m
Breadth (b)	20 m	81 m
Length of shoreline (L)	78 m	469 m
Mean depth ($\bar{z}$)	5.1 m	4.3 m
Shore development (D_L)	1.03	1.19
Mean slope (ϵ)	49°25'	8°51'
Relative depth (z_r, %)	0.45	0.07
$\bar{z}/z_m$	0.46	0.48

(*) These do not take into account the two central areas of Vila basins, occupied by sediment in suspension. As can be observed in figure 1(b), maximum depths at Vila are actually 33 m and 18 m. The addition of the two cones filled with sediment does not significantly alter the parameters shown in this table.

bacteria: only four species (one purple and three green sulphur bacteria) which account for more than 95% of the total bacterial population. Table 3 shows their major morphological and physiological features.

In the monimolimnion of Vila dense populations of green sulfur bacteria are permanently present during the year. Most of them are *Chlorobium phaeobacteroides*, with a maximum of 1.6×10^7 cells/ml at 5 m in July. Also during the summer, in the upper part of the monimolimnion the purple bacterium *Chromatium minus* is present in high numbers (up to 5.5×10^5 cells/ml, at 5 m). Total numbers of *Chlorobium* spp. and *Chromatium minus* in the water column of Vila during two days of summer and two days of winter sampling are shown in Table 4.

Bacterial numbers during the year in Siso are indicated in the accompanying paper (Abella *et al.*, 1980). *Chlorobium* spp. total numbers remain at the same level as those in Vila, with a maximum of about 1.2×10^7 cells/ml, but in Siso the maxima are reached in winter time (March and December) instead of in the summer, as is the case in Vila. On the other hand, the maximum total numbers of *Chromatium* are larger than in Vila (up to 2.5×10^6 cells/ml are found near the surface from May to September). Those higher population densities are related to the conditions prevailing in Siso (sulfide almost up to the surface and hence a better light environment), which allows the purple sulfur bacteria to thrive in the upper layers.

In spite of the fact that it is not the object of this paper, it is worth mentioning that in the anaerobic sediment (located in the Siso and most of the Vila basins) high numbers of *Desulfovibrio desulfuricans* are detected. Sulfate reduction by this bacterium is responsible for the high concentrations of H_2S present in both lakes.

Table 2. Major physical and chemical parameters of Siso and Vila lakes in summer (30.7.78).

Water body	Depth (m)	Temperature (°C)	Conductivity (μmhos/cm at 20°C)	Alkalinity (meq/l)	pH	Redox (mV)	SO_4^{2-} (mg/l)	H_2S (mg/l)	O_2 (mg/l)	NO_3^- (mg/l)	NH_4^+ (mg/l)	PO_4^{3-} (mg/l)
Siso	0	21.4	2080	5.0	7.5	− 31	715	0	5.3	0.03	5.4	0.10
	1	18.5	2090	5.0	7.4	−246	984	88	− (a)	−	23.0	−
	3	15.8	2105	5.3	7.4	−335	960	97	−	−	22.5	−
	5	15.5	2120	6.5	7.4	−335	891	100	−	−	25.2	−
	7	15.0	2128	6.3	7.4	−349	920	130	−	−	25.0	−
	10	15.0	2130	6.7	7.4	−349	945	152	−	−	32.8	−
Vila	0	22.0	1100	4.3	8.0	+270	560	0	9.0	0.32	2.2	0.03
	2	21.5	1100	4.4	8.0	+270	560	0	9.5	0.30	2.7	0.02
	4	19.4	2100	7.7	7.6	−102	900	11	−	−	8.5	−
	6	15.1	2450	11.5	7.5	−340	1230	248	−	−	35.0	−
	8	14.0	2450	11.4	7.5	−390	1240	246	−	−	36.0	−
	below 9 (b)	17.0	2900	* (c)	7.5	−200	*	*	−	*	*	*

(a) −: undetectable

(b) central holes of each basin, filled with sediment in suspension (average density of 1.35 g/cm³), some of whose parameters are shown here.

(c) *: not measured

165

Table 3. Major morphological and physiological characteristics of purple and green sulfur bacteria from Siso and Vila lakes.

Species	Size[a] Length(μm)	Width(μm)	Volume (μm³)	Carotenoid	Major bacterio-chlorophyll	Color of pure culture	Motility[b]
Chromatium minus	2.83±0.31	2.27±0.26	8.39	Okenone	Bchl a	Pink	+
Clorobium phaeobacteroides	2.01±0.35	0.61±0.17	0.53	Isorenieratene	Bchl d	Brown	-
Chlorobium phaeovibrioides	1.84±0.28	0.43±0.14	0.25	Isorenieratene	Bchl d	Brown	-
Chlorobium limicola	2.53±0.30	0.76±0.11	1.03	Chlorobactene	Bchl c	Green	-

(a) Size calculated by electron microscopy of field samples. Confidence level 99%
(b) +: motility by means of polar flagella; -: unmotile.

Table 4. Total numbers of *Chromatium* and *Chlorobium* in the water column of Vila lake at two different times of the year. Sulfide concentrations at the same depths are also shown.

Depth (m)	Summer (15.7.78) Chromatium (cells/ml x10⁵)	Chlorobium (cells/ml x10⁶)	H₂S (mg/l)	Winter (16.12.78) Chromatium (cells/ml x10⁵)	Chlorobium (cells/ml x10⁶)	H₂S (mg/l)
4	0	0	0	0	0	0
4.5	2.60	11.00	10.9	0	0	0
5	5.40	16.00	42.0	0	0	0.1
5.5	1.10	5.70	115.6	1.03	4.00	98.0
6	1.00	3.10	180.0	0.51	5.50	106.0
7	0.90	2.90	191.3	0.13	4.2	118.0

Vertical distribution of bacterial populations. Effects on water color, light penetration and turbidity

Figure 2 shows the evolution through the year of the populations maxima (below) for *Chromatium* and *Chlorobium* in relation to the isopleths of sulfide concentration and water transparency (Secchi disc readings). *Chromatium* maxima in each lake adjust to the levels which the sulfide concentrations permit (<60 mg/l).

The values of the *Chromatium* population in Vila, presented in Fig. 2, show total numbers higher than 10^5 cells/ml, above which level the water acquires a pink tint, due to the large volume of *Chromatium* in comparison to that of *Chlorobium* (see Table 3). In Siso, the values of *Chromatium* reach over 10^6 cells/ml; in this case the water has an intense pink color. Also illustrated are values for *Chlorobium* in both lakes above 10^6 cells/ml; the water looks brownish-green when the *Chromatium* population is lower than 10^5 cells/ml (Guerrero & Abella, 1978).

Turbidity (expressed as the percentage of transmittance at 480 nm at the different depths) and light penetration (percentage of incident light) in the water column are shown in Fig. 3. In Vila,

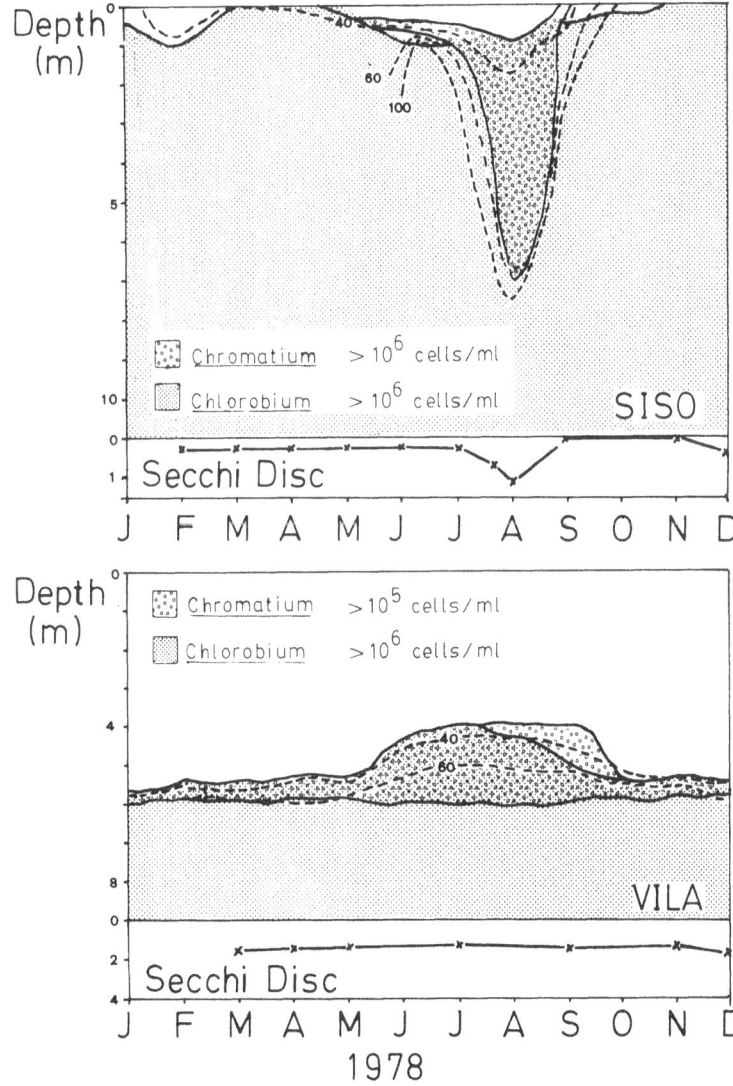

Fig. 2. Distribution through the year of maximum total numbers of *Chromatium minus* and *Chlorobium* spp. in Siso and Vila lakes in relation to Secchi disc readings and sulfide concentration. (Broken lines indicate isopleths of H$_2$S in mg/l.)

turbidity and light penetration profiles show little variation during the year. The monimolimnion shows a sharp attenuation ($\eta = 5.56$ m^{-1}) and turbidity ($T_{480} = 30\%$) between 4.5–6 m depth, where the bacterial populations are very high (see Table 4). In contrast, Siso presents two distinct patterns in the optical properties of the water. During the mixing period, light attenuation is constant ($\eta = 1.45$ m^{-1}) as is turbidity ($T_{480} = 72\%$). During the summer stratification, the metalimnion (between 1–3 m) shows an intense light attenuation and turbidity ($\eta = 4.85$ m^{-1}, $T_{480} = 10\%$) due to the high concentrations of *Chromatium* and *Chlorobium* populations. Below the 3.5 m depth, turbidity is very low ($T_{480} = 90\%$).

Specific pigment content variation

Table 5 presents the mean values of the whole water column of the specific pigment content and

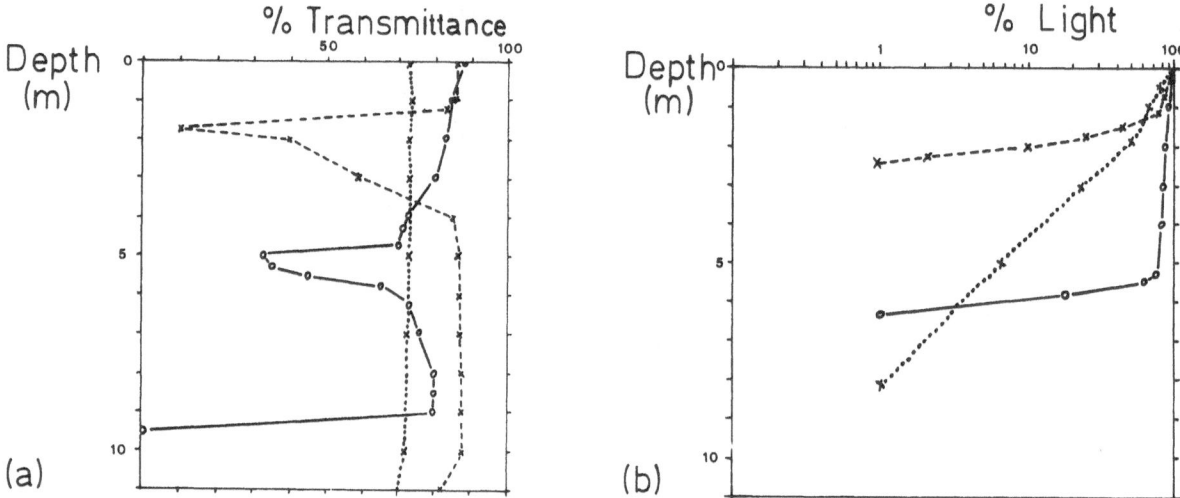

Fig. 3. Turbidity and light penetration is Siso and Vila lakes. (a) Turbidity (% transmittance) at different depths of Siso during winter (×·····×) and summer (×----×) and Vila through the year (O——O). (b) Light penetration (% of incident light) in the water column of Siso during winter (×·····×) and summer (×---×) and in Vila through the year (O——O).

Table 5. Comparison of the water column values for the specific pigment contents and carotenoid/bacteriochlorophyll ratios between summer and winter populations of *Chlorobium* and *Chromatium* from Siso and Vila lakes.

Water body	Season	Chlorobium			Chromatium		
		μg bchl.d / mg w.w.	A.U.chl.+isr. / mg w.w.	A.U.chl.+isr. / μg bchl. d	μg bchl.a / mg w.w.	A.U.okn. / mg w.w.	A.U.okn. / μg bhcl.a
Siso	Summer*	90.0	20.0	0.22	6.0	0.2	0.03
	Winter**	3.0	0.3	0.10	4.0	0.8	0.20
Vila	Summer*	13.6	19.9	1.50	4.9	4.8	0.99
	Winter**	1.0	0.2	0.23	6.4	0.1	0.01

Abreviations: chl.+isr., chlorobactene+isorenieratene ; okn., okenone .

* Sampling of 15.7.78

** Sampling of 16.12.78

carotenoid/bacteriochlorophyll ratio, corresponding to one day of summer (15.7.78) and one day of winter (16.12.78) sampling. The results show that *Chlorobium* populations in both lakes have much higher specific pigment contents in summer than in winter (15–30 times more bchl. *d* and 70–100 times more chlorobactene + isorenieratene). However, *Chromatium* maintains its specific content of bchl. *a*, although with some differences in okenone in Vila. The carotenoid/bacteriochlorophyll ratios ot purple and green bacteria in both lakes are generally higher in summer than in winter. The only exception is the lower ratio for *Chromatium* in summer in Siso, which could be explained by the fact that this population is located near the surface, with more available light.

Discussion

In the anaerobic zones of Vila and Siso lakes, the respective communities of photosynthetic sulfur bacteria (*Chromatium minus* and *Chlorobium* spp.) show, throughout the year, differences in total numbers, locations in the water column and specific pigment content.

The presence of photosynthetic sulfur bacteria is initially determined by the entrance of sulfated water from the bottom. The sulfate inflow and the deposition of organic matter give rise to a "sulfuretum," where the heterotrophic *Desulfovibrio*, in the sediment, uses sulfate as an electron acceptor and produces sulfide, which, in turn, is utilized by the photosynthetic sulfur bacteria, in the water, as an electron donor, producing sulfur and sulfate. Sedimentation of the planktonic photosynthetic bacteria provides organic matter for *Desulfovibrio* and permits reinitiation of the cycle.

In Vila and Siso lakes the water inflow, the turnover time and the morphometry of the basins account for the presence and magnitude of the different physico-chemical parameters and consequently determine the development of the photosynthetic bacterial populations. Those populations growing in the mónimolimnion of Vila possess very stable physico-chemical conditions and a good light climate due to the fact that the chemocline is permanent through the year and located

from 4 to 5 m depth. On the other hand, the populations of Siso are periodically stressed by the fall overturn and destruction of the summer stratification.

Therefore, in the studied lakes two characteristic vertical distributions can be observed. The first (type I), shows a uniform vertical distribution of total numbers, photosynthetic pigments and physico-chemical parameters, due to the mixing process. The second (type II), is a consequence of the sharp gradients of the cited parameters due to thermal or chemical stratification.

In Siso (see Fig. 3) both types of distribution are found. Type I is present during most of the year (holomixis); *Chromatium* and *Chlorobium* populations are mixed and more or less uniformly located in the water column. Type II is found during the summer stratification, when *Chromatium* tends to accumulate in the upper part of the water column, with total numbers larger than 10^6 cells/ml, between the minimum and maximum sulfide concentrations (5 and 60 mg/l). *Chromatium* form a thick plate and have a shading effect on the *Chlorobium* population (Abella *et al.*, 1980), the latter showing its lowest total annual numbers. On the other hand, in Vila only the type II distribution is found through the year: *Chromatium* forms a plate (with maximum numbers of the order of 10^5 cells/ml) in the chemocline and has a permanent shading effect on the *Chlorobium* population. Nevertheless, the thickness of that plate varies during the year, only being wider than 0.5 m from June to September (see Fig. 2). The lower maximum total numbers to which *Chromatium* grows in Vila, in comparison with Siso, could be explained because of the attenuation of light in the water column (4–5 m) and the large algal population of the mixolimnion.

Besides the total numbers of bacteria, it is necessary to take into account the variation in the photosynthetic pigments with respect to biomass (specific content) and the relative amount of careteniods and bacteriochlorophylls in the population.

In Fig. 4, representative profiles of the visible radiation reaching the different depths in Siso and Vila lakes in winter and summer are compared to the pigment content of *Chromatium* and *Chlorobium* in the water column. Figure 4 (bot-

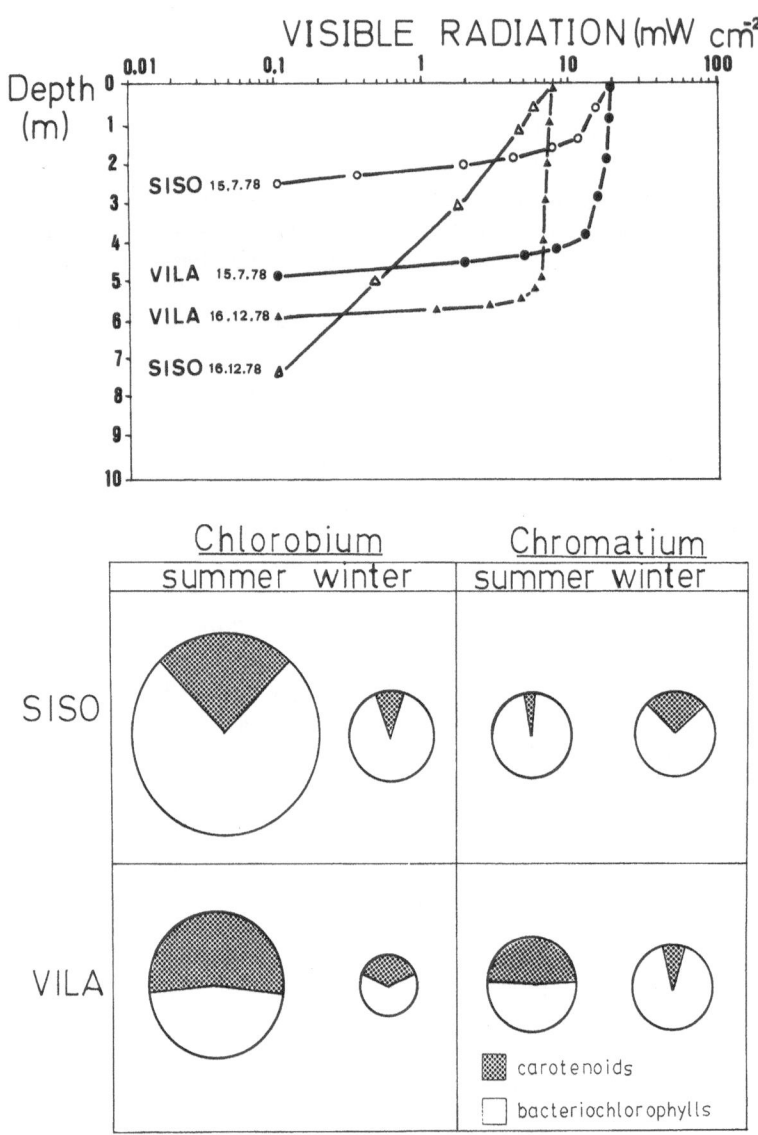

Fig. 4. Top: Vertical profiles of visible radiation in Siso and Vila lakes during one sampling day of summer (15.7.78) and winter (16.12.78). Visible radiation at the surface was estimated as the 45% of the solarimeter measurement (total radiation); radiation at each depth was calculated from the percentage of light penetration (see Fig. 3b). Bottom: Representation of the photosynthetic pigment values in the water column for *Chlorobium* and *Chromatium* at the above mentioned dates. The area of each circle is proportional to the square root of the total specific pigment content. (For absolute values of carotenoids and bacteriochlorophylls, see Table 5.)

tom) schematizes the variation of the photosynthetic pigments of *Chlorobium* and *Chromatium* using the data of Table 5. The area of each circle is proportional to the square root of the total specific pigment content. Figure 4 (top) indicates the amount of visible radiation in mW cm^{-2} at every depth.

During the summer in Siso lake, *Chlorobium*

significantly increases its specific content of bacteriochlorophylls and carotenoids, as well as the carotenoid/bacteriochlorophyll ratio. This increase with respect to the winter values is caused by the shading of the *Chromatium* plate, which is located above the *Chlorobium* population. This effect is corroborated by the diminution of available light energy below 2 m depth (see Fig. 4, top). In contrast, the *Chromatium* population has rather similar total specific contents during the year, only it shows differences between its carotenoid/bacteriochlorophyll ratios: the lower values in summer are probably due to its location at 1–2 m depth and to the high transparency of the layers above.

In Vila lake, the *Chlorobium* population also increases its specific pigment content and carotenoid/bacteriochlorophyll ratio during the summer. There are no important differences in the available radiation observed in the layers with the largest population densities (4.5–5 m in summer and 5.5–6 m in winter, see Table 4); this situation could be explained by the combined filtering effect of the algal (in the mixolimnion) and *Chromatium* populations, both of which are much higher in summer than in winter. In its turn, *Chromatium* shows similar total pigment content in summer as in winter, but increases the pigments ratio in favor of carotenoids, possibly because of the selective filtering of light due to the algae, which enriches the blue-green part of the spectrum, where the maximum absorption of okenone is found.

The energy profiles of light are strongly influenced by the vertical distribution of both populations of photosynthetic sulfur bacteria. For this reason, and taking into account that in summer the *Chromatium* plate is denser and thicker in both lakes, it is possible to explain the relative increase in specific pigment content of *Chlorobium* independently of the fact that in summer there is more radiant energy reaching the water surface.

The results of this field work agree with different laboratory studies. Holt & Marr (1965), Takahashi *et al.* (1972) and Broch–Due *et al.* (1978) showed that photosynthetic bacteria adapt to low illumination by increasing the specific bacteriochlorophyll content. With respect to the carotenoids, our data are not in accordance with the available literature. Takahashi *et al.* (1972) and Broch-Due *et al.* (1978) find a parallel decrease between specific carotenoid content and light intensity. At present, it is not possible to compare directly field results with those of the laboratory, because of the different light conditions provided by incandescent and fluorescent lights. Probably the quality of light affects

differentially the biosynthesis of carotenoids and bacteriochlorophylls. We have in progress further field studies on the way in which spectral quality may affect variation in the carotene and bacteriochlorophyll content of photosynthetic bacteria. It is hoped this will lead to a better understanding of the adaptations to changing conditions in their natural environment.

References

Abella, C., Montesinos, E. & Guerrero, R. 1980. Field studies on the dynamics of competition for available light between purple and green sulfur bacteria. (This volume.)

Bergstein, T., Henis, Y. & Cavari, B. Z. 1979. Investigations on the photosynthetic sulfur bacterium Chlorobium phaeobacteroides causing seasonal blooms in Lake Kinneret. Can. J. Microbiol. 25: 999–1007.

Broch-Due, M., Ormerod, J. G. & Fjerdingen, B. S. 1978. Effect of light intensity on vesicle formation in Chlorobium. Arch. Microbiol. 116: 269–274.

Golterman, H. L., Stringer, R. & Ohnstad, M. A. M. (eds). 1978. Methods for physical and chemical analysis of fresh water. Blackwell, 2nd ed.

Guerrero, R., & Abella, C. 1978. Dinámica espaciotemporal de las poblaciones bacterianas fotosintéticas en una laguna anaerobia de aguas sulfurosas. Oecol. aquatica 3: 193–205.

Holt, C. S. & Marr, A. G. 1965. Effect of light intensity on the formation of intracytoplasmic membrane in Rhodospirillum rubrum. J. Bacteriol. 83: 1421–1429.

Hutchinson, G. E. 1957. A treatise on limnology. Vol. I. J. Wiley.

Kuznetsov, S. I. 1977. Trends in the development of ecological microbiology. *In.* Droop. M. R. & H. W. Jannash (eds.). Advances in aquatic microbiology, pp. 1–48. Academic Press.

Parsons, T. R. & Strickland, J. D. H. 1963. Discussion on spectrophotometric determination of marine-plant pigments, with revised equations for ascertaining chlorophylls and carotenoids. J. Mar. Res. 21: 155–163.

Pfennig, N. 1967. Photosynthetic bacteria. Annu. Rev. Microbiol. 21: 285–324.

Takahashi, M. & Ichimura, S. 1968. Vertical distribution and organic matter production of photosynthetic sulfur bacteria in Japanese lakes. Limnol. Oceanogr. 13: 644–655.

Takahashi, M. & Ichimura, S. 1970. Photosynthetic properties and growth of photosynthetic sulfur bacteria in lakes. Limnol. Oceanogr. 15: 929–944.

Takahashi, M., Shiokawa, K. & Ichimura, S. 1972. Photosynthetic characteristic of a purple sulfur bacterium grown under different light intensities. Can. J. Microbiol. 18: 1825–1828.

Trüper, H. G. & Genovese, S. 1968. Characterization of photosynthetic sulfur bacteria causing red water in lake Faro (Messina, Sicily). Limnol. Oceanogr. 13: 225–232.

FIELD STUDIES ON THE COMPETITION BETWEEN PURPLE AND GREEN SULFUR BACTERIA FOR AVAILABLE LIGHT (LAKE SISO, SPAIN)

C. ABELLA, E. MONTESINOS & R. GUERRERO

Abstract

Photosynthetic sulfur bacteria (*Chromatiaceae* and *Chlorobiaceae*) living in the same water body compete for available light. In Siso lake, a small anoxic holomictic lake near to Banyoles (Girona, Spain), the competitive balance changes seasonally depending on physicochemical gradients and light. Maximum total numbers of both groups of bacteria are complementary throughout the year. In summer time, the presence of motile *Chromatium* plates near the surface act as a biological light filter on the *Chlorobium* populations growing below. The latter adapt to the modified light spectrum and intensity by increasing both the specific pigment content and the carotenoid/bacteriochlorophyll ratio.

Introduction

Few field studies have been done on the competition between purple and green sulfur bacteria for available light (Trüpper & Genovese, 1968; Matheron & Baulaigue, 1977), mainly due to the few situations which exist where the two populations thrive in the same water body. But even rarer are studies that follow both populations throughout the year (Genovese et al., 1963).

The present paper describes the annual changes in total numbers and photosynthetic pigments of a mixed population of purple and green sulfur bacteria in an anaerobic lake. This approach can explain the different strategies of these bacteria under variable light climates in natural conditions. The study was made in Siso lake, a small anoxic water body (26 m long, 20 m wide, 11 m maximum depth) located near Banyoles Lake (Girona, Spain), whose physico-chemical characteristics have been described elsewhere (Guerrero & Abella, 1978). We could undertake this study because of the low species diversity (only two dominants), which simplifies the work and allows us to treat the population like a two member system. Provided that we know the relative composition of the photosynthetic bacterial population by means of cell morphology, bacteriochlorophyll, and carotenoid content, it is then possible to study the changes, of the specific pigment content, in relation to the available light throughout the year.

Material and methods

Total bacterial numbers were determined by direct microscopy on membrane filters (Sartorius, mean pore size 0.45μm), stained following Lumpkins & Arveson (1968) method, with acid fuchsin and methylene blue dyes. Bacteria show, after staining, a pink-red color against a light blue background. *Chromatium* cells are big ovals, clearly distinguishable from the much smaller and bacillary *Chlorobium* cells.

Photosynthetic pigments were extracted from pure cultures of bacteria originally isolated from Siso and from field samples of the same water body. They were separated by thin layer and column chromatography with aluminium oxide (Jensen et al., 1964, Liaaen–Jensen et al., 1964) and identified by their absorption spectra in petroleum ether.

The *in vivo* absorption spectra were measured both in pure culture and field samples. Bacteria were separated by filtration, sonified and centrifuged at $15\,000 \times g$ during 15 min. The supernatant, which contains the photosynthetic membranes, was scanned with a Pye Unicam 1700 spectrophotometer, from 350 to 850 nm. Other methods concerning biomass determination and pigment quantification are described in the previous paper (Guerrero *et al.*, 1980).

Results

Species identification.

Four species of photosynthetic sulfur bacteria were isolated from Siso lake on agar shake tubes with Pfennig (1965) medium. The only purple bacteria present in high numbers was *Chromatium minus*. It was identified by electron microscopy (ovals 2.3 μm wide) and pigment composition (bacteriochlorophyll *a* and okenone). The other significant bacteria belong to the green sulfur group and they were identified, also by electron microscopy and pigment composition, as the following three species: (1) *Chlorobium phaeobacteroides* (rodshaped; bacteriochlorophyll *d* and isorenieratene), which constitutes around the year the 94–97% of the green sulfur bacteria. (2) *Chlorobium phaeovibrioides* (curved to vibrioshaped; bacteriochlorophyll *d* and isorenieratene), and (3) *Chlorobium limicola* (rodshaped; bacteriochlorophyll *c* and chlorobactene), found only occasionally.

In spite of the fact that in nature bacterial populations are frequently found composed of many species, in the anaerobic waters of Siso the diversity is very low (as is the case in other similar habitats) and the four species mentioned above account for more than 95% of the total aquatic bacterial population.

Annual bacterial numbers in the water column

In field studies, total counts are a good indicator of population growth because they reflect the variation of the physicochemical conditions which favor the increase of one or another component of the mixed population. Figure 1 presents the total numbers of *Chromatium* (*C. minus*) and *Chlorobium* (the three mentioned species) in the water column of Siso lake at different times of the year.

In winter and early spring *Chromatium* (Fig. 1a) is present in low numbers ($2-3 \times 10^5$ cells/ml) and is evenly distributed throughout the water column. In May and June it can be observed as a typical plate at 1–2 m depth, with a maxima of $2-3 \times 10^6$ cells/ml. This plate always coincides with the onset of thermal stratification of the water column just below the thermocline. From July to September the population is rather higher ($>10^6$ cells/ml) and more or less uniformly distributed with depth.

The spatial-temporal distribution of *Chlorobium* (Fig. 1b) is almost complementary to that of *Chromatium*. The maximum of total counts is found in winter ($>10^7$ cells/ml) and the minimum in summer (2×10^6 cells/ml), which is just the opposite case to *Chromatium*. It is interesting to note that in Siso lake the photosynthetic bacterial community reaches the surface during almost the whole year, excluding late summer when an oxygenated epilimnion 1–2 m deep is formed.

Photosynthetic pigments

The absorption spectra of pure *Chromatium* and *Chlorobium* cultures as well as water samples from Siso are shown in Fig. 2.

The only purple bacteria, *Chromatium minus*, present had three main peaks both *in vivo* (sonicated cells) and in acetone extracts. The *in vivo* peaks are at 375 and 830 nm, due to bacteriochlorophyll (bchl.) *a* and at 510–540 nm, due to okenone (Fig. 2a). The acetone extract peaks are at 365 and 773 nm, due to bchl. *a*, and at 490 nm, due to okenone (Fig. 2b).

The *in vivo* extracts of pure cultures of *Chlorobium phaeobacteroides* show a peak at 725 nm, due to bchl. *d* and another at 460–520 nm, due to isorenieratene (Fig. 2a). The acetone extract peaks are at 654 nm, due to bchl. *d*, and at 470 nm, due to isorenieratene (Fig. 2b).

Table 1 presents the main absorption peaks of the cellular pigments in Siso water samples as well as the pigments of the same samples previously separated by thin layer chromatography and identified in two different solvents: acetone and petroleum ether. The data confirm the species

174

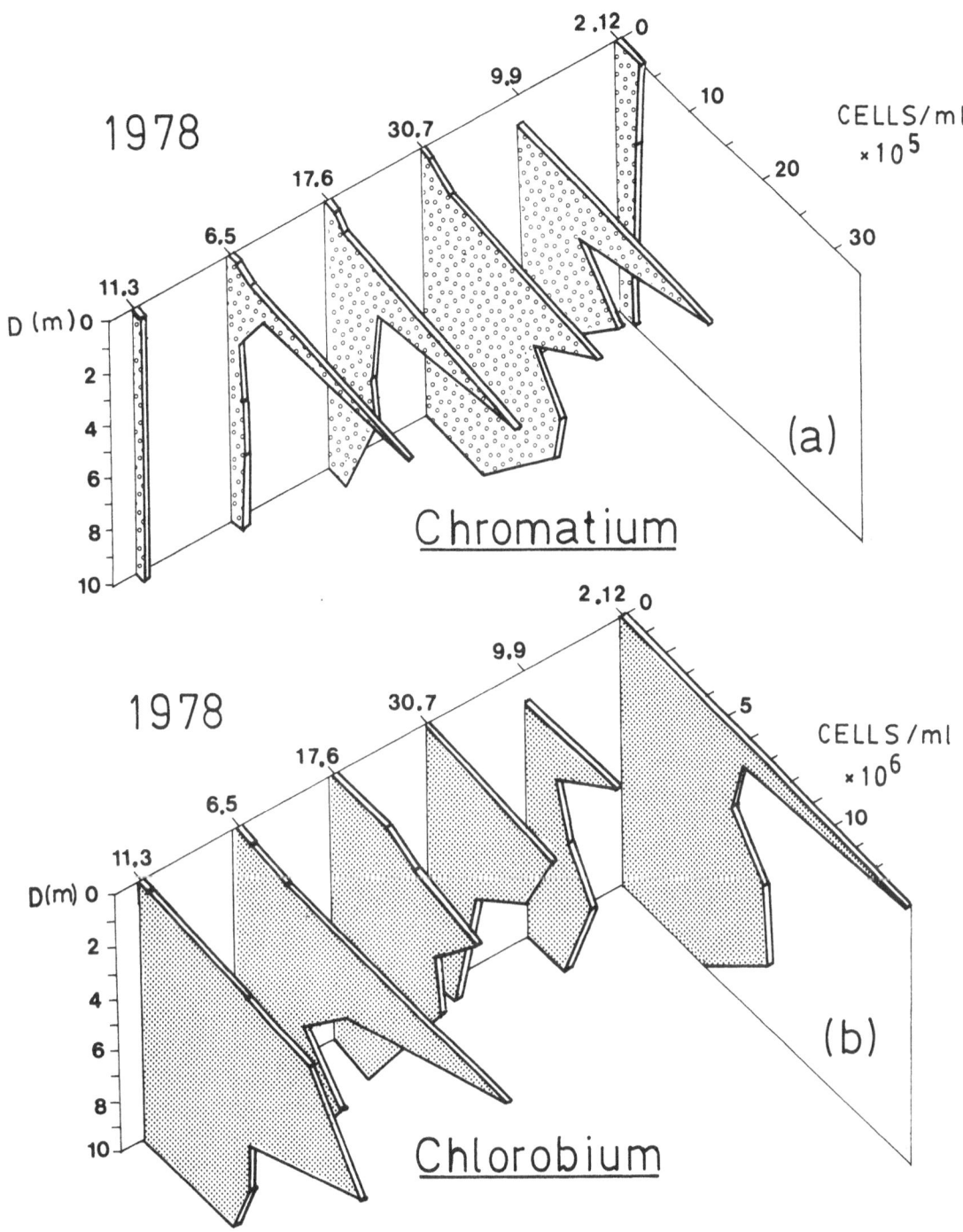

Fig. 1. Seasonal changes in the vertical distribution of total cell numbers of *Chromatium* (a) and *Chlorobium* (b) during 1978 in Siso lake.

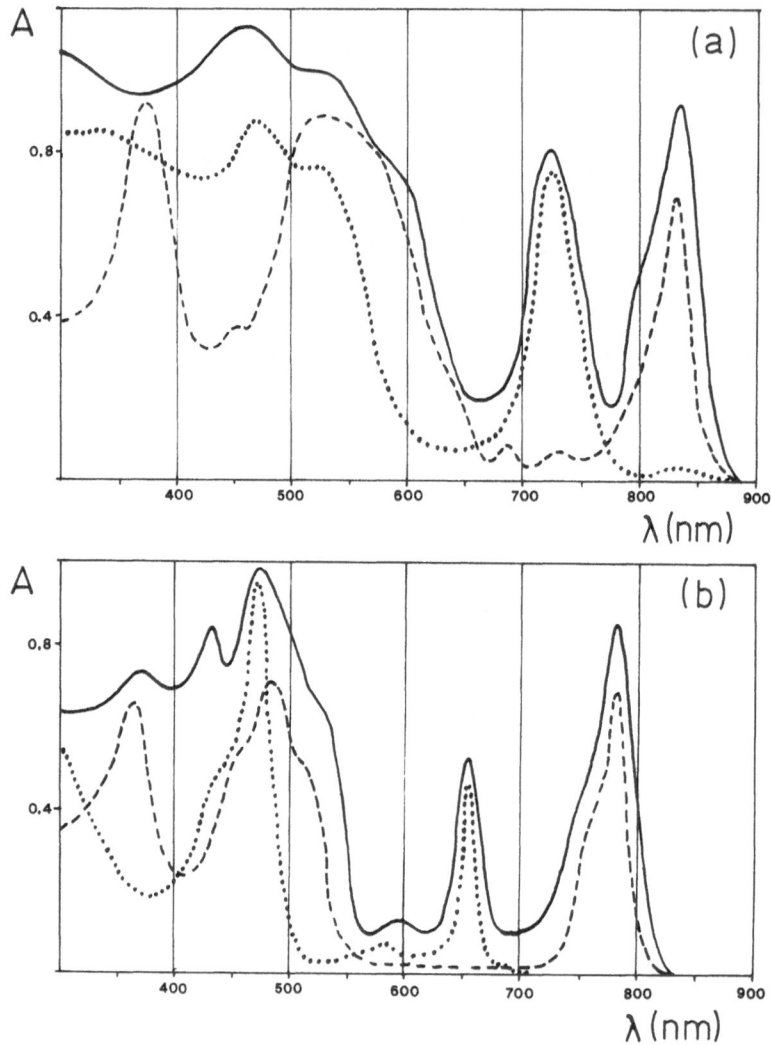

Fig. 2. Absorption spectra of *Chromatium* (– – –) and *Chlorobium* (· · · · ·) pure cultures and of water samples from Siso lake (——). (a) Sonicated cells. (b) Crude extracts in acetone 90%.

identification mentioned above and allow us to quantify both the specific content in bacterio-chlorophylls [μg/mg wet weight (w.w.)] and carotenoids [arbitrary units (A.U.)/mg w.w.] using the absorption maxima (Guerrero *et al.*, 1980).

Specific content of photosynthetic pigments

Figure 3 presents the mean values in the water column, during the year, of three parameters belonging to *Chromatium minus:* (1) specific content of okenone, (2) bchl. *a*, and (3) total cells number per milliliter.

Figure 4 shows the same mean values for the corresponding parameters of *Chlorobium* spp., namely, (1) isorenieratene plus chlorobactene, (2) bchl. *d*, and (3) cells/ml.

The population of *Chromatium minus* shows four characteristic phases. During the lag phase (January to April) the specific content of okenone is high (maximum of 0.80 A.U./mg w.w., in March). During the growing phase (April to June) okenone diminishes to values of 0.20 A.U./mg w.w. During the balanced phase (July to mid September) okenone starts to increase. Finally, since mid September the population declines very steeply and in a month the

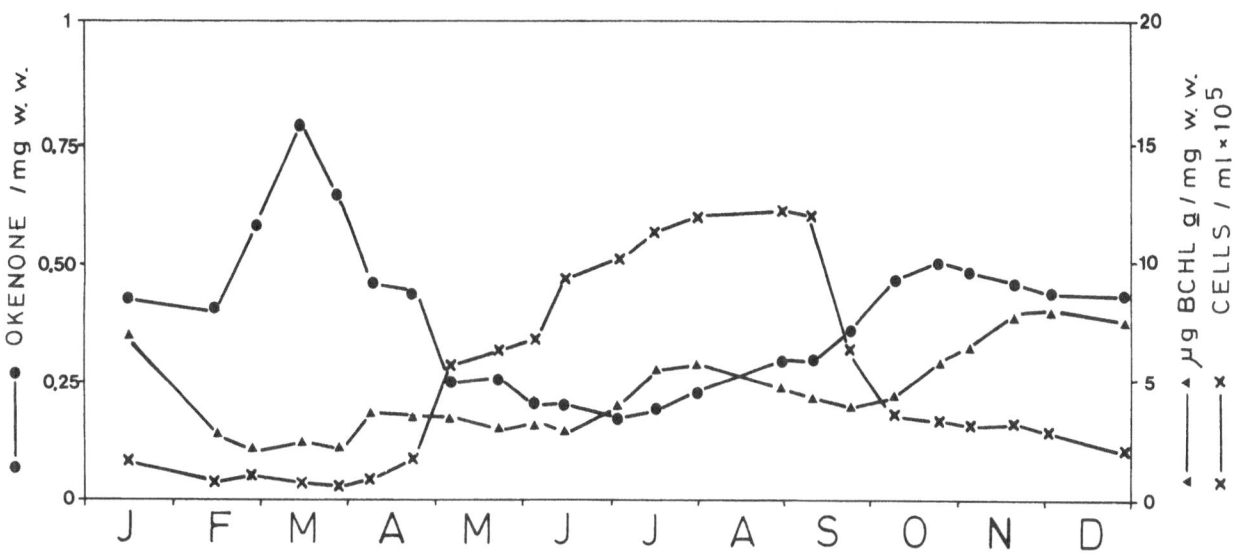

Fig. 3. Changes in the mean values in the water column of Siso lake of total number and specific content of okenone and bacteriochlorophyll *a* in *Chromatium* during 1978.

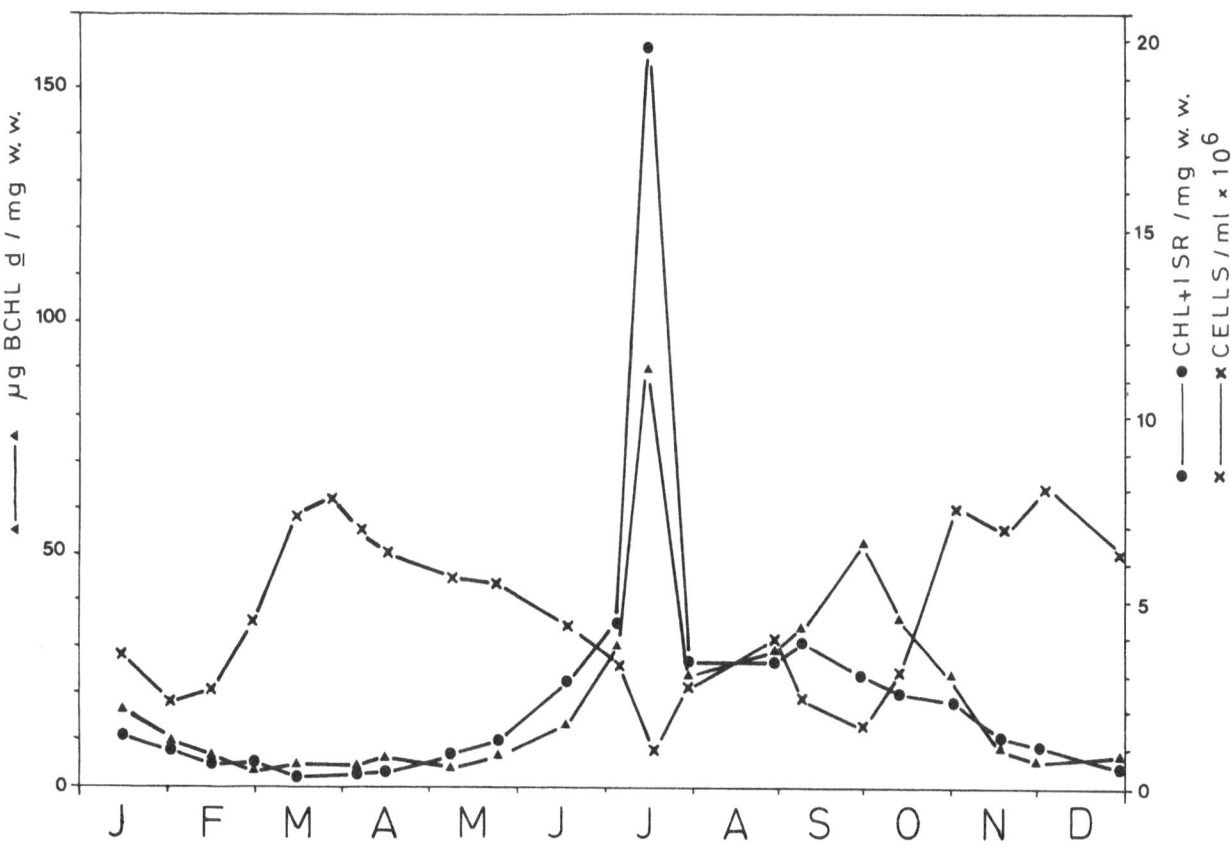

Fig. 4. Changes in the mean values in the water column of Siso lake of the total number and specific content of chlorobactene plus isorenieratene and bacteriochlorophyll *d* in *Chlorobium* spp. during 1978.

Table 1. Main absorption maxima of extracts of Siso water samples and of photosynthetic pigments of the same samples previously separated by thin layer chromatography (Data in nm).

Pigment	Water samples (λmax. in acetone)		Chromatography λmax. in acetone	λmax. in ether
Bacterio-chlorophyll a	365	773	360 580 773	-
Bacterio-chlorophyll d	610	654	410 428 608 654	-
Okenone		520	-	458 482 512
Isorenieratene		470	415 444 475	426 450 478
Chlorobactene		470	435 468	434 462

(-) Not determined.

specific content of okenone reaches the values of the former lag phase. The specific content of bchl. *a* remains rather constant throughout the year. (See Fig. 3.)

On the other hand, the population of *Chlorobium* spp. shows the maxima, in March and in October–December, precisely when *Chromatium* population is lowest. The specific content of carotenoids (isorenieratene plus chlorobactene) and bchl. *d* is the highest (maxima of 20 A.U./mg w.w. and 90 μg/mg w.w., respectively), when the population density is the lowest (in July). (See Fig. 4.)

Discussion

To study the competition of purple and green bacteria in nature is necessary to find optimal conditions (sulfide concentration and light climate) for both populations in the same water body. There are few cases in the literature where this situation is fulfilled. Cohen *et al.* (1977) found two separated plates of *Chromatium* and *Prosthecochloris* in Solar Lake, but the populations disappeared in two-three months due to the presence of aerobic conditions in the whole water column. Gorlenko

& Leveba (1971) have also described the existence of two distinct plates of photosynthetic sulfur bacteria in Lake Kononier. In fact, clear cut plates are not permanent but, instead, purple and green sulfur bacteria are present during most of the year in mixed layers, as is the case in Siso (Guerrero & Abella, 1978) and Faro (Trüpper & Genovese, 1968; Sorokin & Donato, 1975).

In Siso lake cyclical changes in the spatial distribution of purple and green sulfur bacteria are found throughout the year. These changes enable the study of competition between these populations. The interpretation of field results could be facilitated by studying the seasonal changes in physicochemical gradients which influence their population dynamics.

Thus, this paper tries to emphasize the importance of field studies in order to understand laboratory findings. It has the advantage of studying the populations in their natural underwater light conditions, which are difficult to obtain in the laboratory. However, it is rather hard to interpret the population dynamics for all depths in relation to the light climate, and so mean values for the whole water column are used.

Chromatium seasonal changes during 1978 (see

Figs. 1a and 3) are linked to the sulfide, light and temperature conditions, with optimum growth during summer. The largest *Chromatium* populations are directly related to thermal stratification in the water column which starts in April and ends in September. It reaches 5–6 m of depth when sulfide concentration is below 60 mg/l (Guerrero *et al.* 1980).

The shading effect of *Chromatium* on *Chlorobium* can be clearly seen in Siso lake (see Figs. 1b and 4). The effect on the *Chlorobium* population is an increase in the total specific pigment content (carotenoids and bacteriochlorophylls) in June–July (Fig. 4) even when the solar irradiance is at its maximum. *Chromatium* populations are only dense near the surface during May and June. The thickness of the plate, estimated with a transmission meter, was 20–30 cm. The population concentration in this plate is only partially reflected in the total numbers as they are calculated from samples taken by a Ruttner bottle, and so a dilution error is introduced due to the length of the sampler (40 cm).

Keeping in mind this sampling limitation, the sharp decrease in the *Chlorobium* pigment values after July could be explained by a redistribution of *Chromatium* cells over the whole water column with a consequent reduction of the shading effect.

In Fig. 5, the mean values of monthly changes of carotenoids/bchl. ratio in *Chromatium* and *Chlorobium* are presented. The values for *Chromatium* are fairly constant (between 0.05 and 0.1) during most of the year, with the exception of the lag phase, when the values are high (0.26 to 0.46). During this phase, the competition for available light favours the *Chlorobium* population, which is still present in high numbers. For *Chlorobium*, the carotenoids/bchl. *d* maximum values are during the growing phase of *Chromatium*, the ratio being from 0.24 to 0.16 (May to July). The rest of the year it is low (0.04–0.09).

The annual changes in total numbers, specific pigment content, and the ratio of carotenoids/ bchl. are mainly explained by the shading effect and the competition for available light, which is schematized in Fig. 6. In the right side of the figure (upper part), the *in vivo* absorption spectrum of *Chromatium* is represented. The dark areas indicate the wavelengths strongly absorbed by water (Jerlov, 1968; Kirk, 1977). This means that the available light for *Chromatium* photosynthesis corresponds to the blue-green part of the spectrum (450–650 nm). The *in vivo* absorption spectrum of *Chlorobium* is shifted to shorter wavelengths and the carotenoids of those bacteria still can absorb the most penetrating light. However, the projection of the maximum absorption of *Chromatium* carotenoids on the *in vivo* absorption spectrum of *Chlorobium* results in a central shaded area. This shading effect changes the quality of the available light for *Chlorobium* photosynthesis.

The competition for available light between *Chromatium* and *Chlorobium* is expressed in two different ways, motility and pigment adaptation. *Chromatium* cells are flagellated and can move to find their optimum light and sulfide conditions in the water column. These optimal conditions are related to their carotenoid pigments, which absorb in the green part of the spectrum (okenone in *Chromatium minus*, at 540 nm) and for this reason they depend on the penetration of this light, provided that the longer wavelengths have been attenuated by the water.

Chromatium usually is found in the upper part of the anoxic layers, and in the case of Siso just below the thermocline. This situation has for *Chromatium* the advantage of an intense sulfide

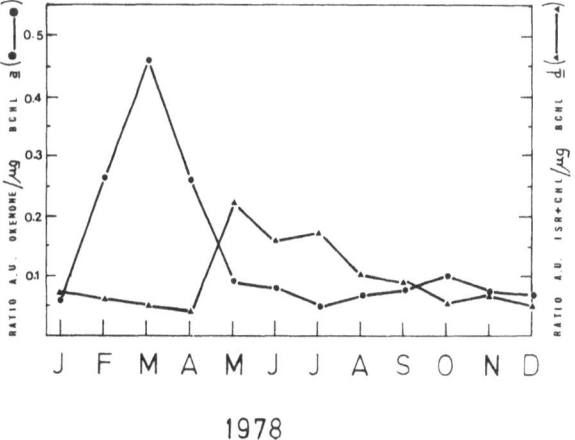

Fig. 5. Ratio between carotenoids and bacteriochlorophylls of *Chromatium* (okenone/bchl. *a*) and *Chlorobium* (isorenieratene plus chlorobactene/bchl. *d*) populations during 1978 in Siso lake. (A.U., arbitrary units of carotenoids.)

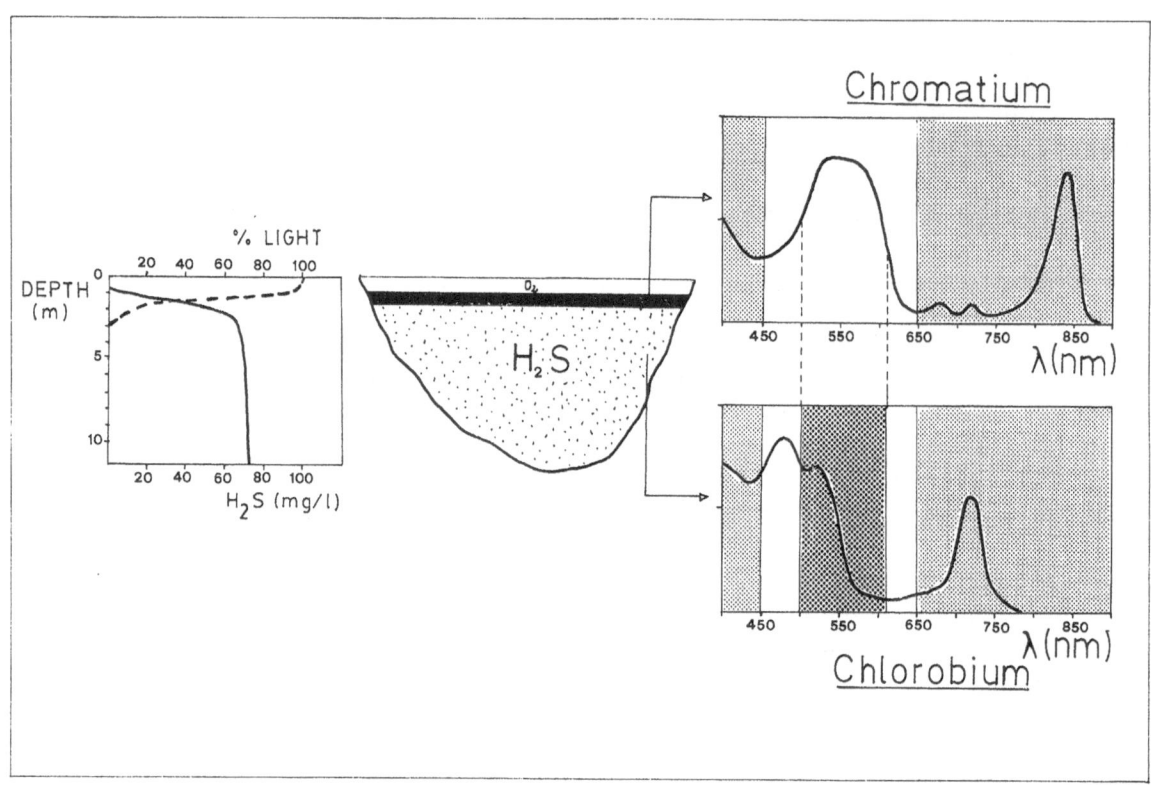

Fig. 6. Diagram of *Chromatium* shading effect on *Chlorobium* and competition for available light in Siso lake during thermal stratification and oxygen until 1 m depth.

gradient even when the bottom sulfide concentration is too high for those bacteria. They accumulate in their optimal sulfide concentration, and can form a plate of 20–30 cm. That plate acts as a biological filter, reducing the intensity and quality of light available for the *Chlorobium* population.

On the other hand, *Chlorobium* is unmotile and tolerates a wide range of sulfide concentrations; hence, they live in the whole water column of Siso. They adapt themselves to the light limiting situation by shifting the absorption of energy to the carotenoids (in this case with an absorption maximum at 520 nm), and increasing the pigment content, as is seen when the *Chromatium* filter is established.

Our conclusions agree with the results of Stephanopoulos & Fredrickson (1979), who studied the coexistence of two phytoplankton populations in the same homogenous environment even though they competed for the available light

spectrum. They concluded that, when light is a limiting growth factor, different photosynthetic populations can coexist in the same environment because of their ability to utilize, differentially, the various wavebands of light which are available to them. In our case it will be necessary, in the future, to measure the actual energy of the light spectrum reaching the different depths during an annual cycle to explain better the behaviour of these populations in nature.

The data presented here stress the importance of biological light filters, resulting from differences in the vertical distribution of the studied populations, on the competition for available light. Purple and green sulfur bacteria adapt to the limiting light conditions increasing their specific pigment content and carotenoid/bacteriochlorophylls quotient. This fact has an important ecological advantage for the studied bacteria because it tends to obviate the effect of competition by

180

means of physiological adaptation, so enabling the coexistence of both populations in the same habitat.

References

Cohen, Y., Krumbein, W. E. & Shilo, M. 1977. Solar Lake (Sinai). 3. Bacterial distribution and production. Limnol. Oceanogr. 22: 621–634.

Genovese, S., Rigano, C. & Macri, C. 1963. Ciclo annuale di osservazioni microbiologiche nel lago di Faro. Atti Soc. Peloritana Sci. Fis., Mat. Nat. 9: 293–329.

Gorlenko, W. M. & Leveba, G. V. 1971. Vertical distribution of photosynthetic bacteria in the Kononier Lake in the Mari, USSR. Microbiologya 40: 651–652.

Guerrero, R. & Abella, C. 1978. Dinámica espaciotemporal de las poblaciones bacterianas fotosintéticas en una laguna anaerobia de aguas sulfurosas. Oecol. aquatica 3: 193–205.

Guerrero, R., Montesinos, E., Esteve, I. & Abella, C. 1980. Physiological adaptation and growth of purple and green sulfur bacteria in a meromictic and an holomictic lakes. (This volume.)

Jensen, A., Aasmundrud, O. & Eimhjellen, K. E. 1964. Chlorophylls of photosynthetic bacteria. Biochim. Biophys. Acta 88: 466–479.

Jerlov, N. J. 1968. Optical oceanography. Elsevier, Amsterdam.

Kirk, J. T. O. 1977. Attenuation of light in natural waters. Aust. J. Mar. Freshwater Res. 28: 497–508.

Liaaen–Jensen, S., Hegge, E., & Jackman, L. M. 1964. Bacterial carotenoids of photosynthetic green bacteria. Acta Chem. Scand. 18: 1703–1718.

Lumpkins, E. D. & Arveson, J. S. 1968. Improved techniques for staining bacteria on membrane filters. Appl. Microbiol. 16: 433–434.

Matheron, R. & Baulaigue, R. 1977. Influence de la pénétration de la lumière solaire sur le développement des bactéries phototrophes sulfureuses dans les environnements marins. Can. J. Microbiol. 23: 267–270.

Pfennig, N. 1965. Anreicherungskulturen für rote und grüne Schwefelbacterien. Zentbl. Bakt. Parasitkde (Abt. 1: Orig) Suppl. 1: 179–189.

Sorokin, J. L. & Donato, N. 1975. On the carbon and sulphur metabolism in the meromictic Lake Faro (Sicily). Hydrobiologia 47: 241–252.

Trüpper, H. G. & Genovese, S. 1968. Characterization of photosynthetic sulfur bacteria causing red water in Lake Faro (Messina, Sicily). Limnol. Oceanogr. 13: 225–232.

Stephanopoulos, G. & Fredrickson, A. G. 1979. Coexistence of photosynthetic microorganisms with growth rates depending on the spectral quality of light. Bull. Math. Biol. 41: 525–542.

ENZYMATIC ACTIVITY OF BACTERIA FROM TWO LAKES OF DIFFERENT TROPHY (LAKE JEZIORAK AND LAKE JASNE, POLAND)

W. DONDERSKI

Abstract

Studies were performed on the occurrence of proteolytic, amylolytic and pectolytic bacteria in water and mud of the eutrophic lake Jeziorak and the mesotrophic lake Jasne. The above bacteria were more numerous in the eutrophic lake. A higher activity of proteolytic enzymes was noted in strains from the eutrophic lake but with higher amylolytic activity in the mesotrophic one. Endopolygalacturonase activity was noted to be higher in strains originating from the mesotrophic lake and of exopolygalacturonases activity in those from the eutrophic lake

Introduction

The source of food for heterotrophic bacteria in water basins is mainly excretions and exudates of plants and animals' remnants of these organisms and organic substances introduced into lakes by sewage and run-off from the surrounding catchment (Henrici, 1939; Kuznetsov, 1951; Collins, 1960; Coler & Gunner, 1969; Nalewajko & Lean, 1972). The identification of the micro-organisms that produce endo- and exoenzymes to decompose different organic compounds as well as an understanding of the factors that affect the synthesis and activity of these enzymes is important for a better understanding of the mechanisms and processes occurring in water basins.

Methods

Studies were carried out with planktonic and benthic bacteria isolated in spring, summer, autumn and winter from the eutrophic lake Jeziorak and mesotrophic lake Jasne. Both lakes are located in the Mazurian district in the northern part of Poland.

After preliminary studies, bacteria which decompose gelatin, starch and pectin were used for research on enzymatic activity. For this purpose bacteria were grown at 22°C during 24–144 hours on a rotary shaker in liquid media adjusted to pH 7.0. The following media were applied:

For proteolytic activity – peptone proteose (Difco) – 5.0 g, $FeSO_4 \cdot 7H_2O$ – 0.1 g, $(NH_4)_2SO_4$ – 0.1 g, gluconate ferrous – 0.1 g, casein (BDH) – 5.0 g, tap water – 1000 ml.;

For amylolytic activity – $Na_2HPO_4 \cdot 12H_2O$ – 2.8 g, KH_2PO_4 – 1.3 g, $CaCl_2 \cdot 2H_2O$ – 0.1 g, $FeSO_4 \cdot 7H_2O$ – 0.05 g, $MgSO_4 \cdot 7H_2O$ – 0.05 g, casein (BDH) – 4.0 g, yeast extract (Difco) – 0.1 g, starch soluble – 5.0 g, tap water – 1000 ml;

For pectolytic activity – $(NH_4)_2SO_4$ – 2.0 g, $MgSO_4 \cdot 7H_2O$ – 0.2 g, $FeSO_4 \cdot 7H_2O$ – 0.05 g, KCl – 1.0 g, K_2HPO_4 – 5.0 g, KH_2PO_4 – 1.0 g, pectin (Sigma) – 5.0 g, yeast extract (Difco) – 0.2 g, tap water – 1000 ml.

Enzymatic activity was determined in cell-free media. Proteolytic activity was studied with Hammerstein's casein as the substrate. Equivalents of tyrosine liberated were determined according to the method of Schacterle and Pollak (1973). Soluble starch was used as the substrate in studies on amylolytic activity. The reducing groups corresponding to glucose equivalents were determined according to the method of Nelson (1944). In studies on pectinolytic enzymes the activity of endo- and expolygalacturonases were determined. Citrus pectin (Sigma) was used as the substrate. Viscosimetric method was applied for endopolygalacturonase activity determination. In studies on exopolygalacturonase activity the free reducing groups were determined according to the method of Nelson. Equivalents of galacturonic acid were also considered. Studies on the above enzymes were performed at 20, 26, 30 and 37°C and at pH 6.0, 7.0, 8.0 and 9.0.

Results and discussion

The results of the studies are presented in Tables 1–3 and Figs. 1–3. The occurrence of proteolytic, amylolytic and pectinolytic bacteria in water and mud in the eutrophic and mesotrophic lakes is shown in Table 1. It appears from this table that the above groups of bacteria were more numerous in the eutrophic lake. Proteolytic bacteria in both lakes were more numerous among the benthic organisms. On the other hand amylolytic bacteria were more numerous in water of the eutrophic lake and in the mesotrophic one in the mud. A reverse phenomenon was noted with the pectinolytic bacteria.

Table 2 shows the activity of proteolytic and amlolytic enzymes of bacteria isolated from both lakes. It appears from the data contained in this table that the activity was higher in strains obtained from the eutrophic lake than in those from the mesotrophic lake. Benthic bacteria of both lakes showed a higher proteolytic activity than the planktonic organisms. The highest proteolytic activity in strains isolated from the eutrophic lake Jeziorak was found in winter whereas from the mesotrophic lake Jasne it was in spring. It seems, that the reason for this is connected with lack of easily available organic substances in the lake. According to Vallentyne (1957) intensive mineralization processes occurring in the upper parts of the bottom sediments raises the temperature in the sediment and in water closely connected with the sediment. This fact may affect the activity of the enzymes synthesized as well as the intensity of mineralization.

The amylolytic activity of planktonic bacteria was higher in strains obtained from the eutrophic

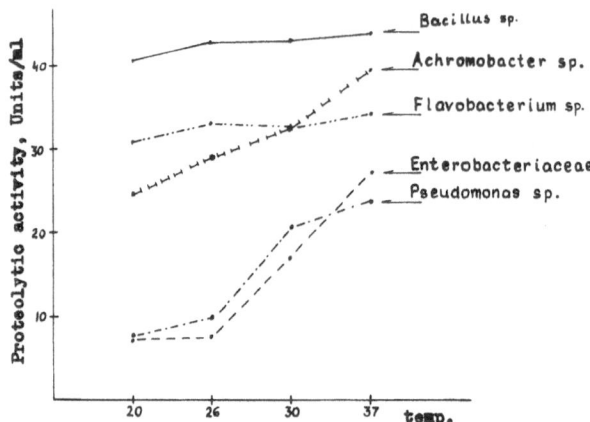

Fig. 1. Activity of the proteolytic enzymes studied at pH 7.0 and different temperatures Explanations: Units/ml = μ moles of product released by 1 ml of cell-free medium/hour.

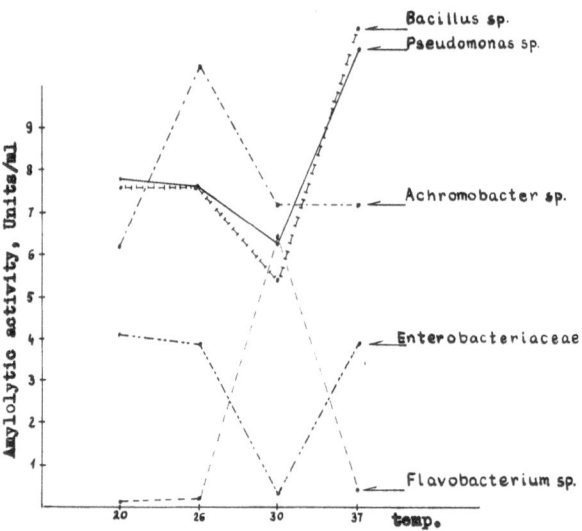

Fig. 2. Activity of amylolytic enzymes studied at pH 7.0 and different temperatures Explanations: see Fig. 1.

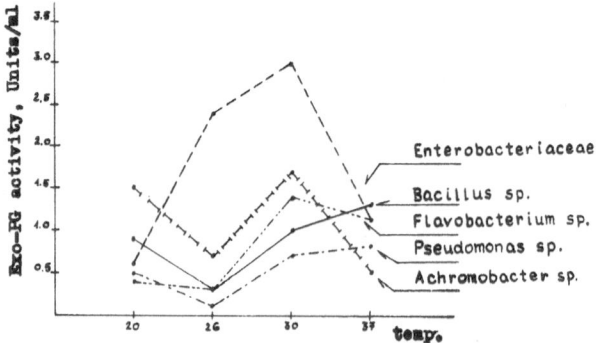

Fig. 3. Activity of the exopolygalacturonase enzymes studied at pH 7.0 and different temperatures Explanations: see Fig. 1.

184

Table 1. Proteolytic, amylolytic and pectinolytic bacteria in water and mud of the lake Jeziorak and Jasne (average values)

Place of sampling		Total number of heterotrophic bacteria	Bacterial types (%)		
			Proteolytic	Amylolytic	Pectinolytic
L. Jeziorak	water	2.680/ml	66	55	6
	mud	87.7×10^5/g d.w.	77	33	22
L. Jasne	water	1.035/ml	50	27	6
	mud	11.3×10^5/g d.w.	60	30	2

Table 2. Proteolytic and amylolytic activity of bacteria isolated from water and mud of the lake Jeziorak and Jasne (average values)

Place of sampling		Number of strains studied	Activity*	
			Proteolytic	Amylolytic
L. Jeziorak	water	40p/20a	62.9 (14.1– 96.3)	84.9 (41.3–165.2)
	mud	40p/20a	111.0 (75.5–162.6)	63.4 (11.2–133.6)
L. Jasne	water	40p/20a	35.2 (3.4– 60.6)	74.1 (14.5–132.6)
	mud	40p/20a	41.6 (6.5– 67.1)	85.4 (46.2–137.1)

Explanations: p = proteolytic bacteria, a = amylolytic bacteria, * = specific activity = μ moles of product released by 1 mg enzymatic proteins per hour

Table 3. Endo- and exo-polygalacturononase activity of bacteria isolated from water and mud of the lake Jeziorak and Jasne (average values)

Place of sampling		Number of strains studied	Activity*	
			endo-PG	exo-PG
L. Jeziorak	water	11	19.5 (4.0–88.4)	136.8 (12.4–242.1)
	mud	20	25.1 (19.2–34.7)	117.2 (36.4–209.4)
L. Jasne	water	8	23.9 (7.9–46.4)	65.2 (15.7–100.3)
	mud	4	21.9 (14.1–29.8)	65.9 (65.2– 66.7)

* See explanations Table 2.

lake. This activity was however higher in benthic bacteria isolated from the mesotrophic lake. In general a higher amylolytic activity was noted in bacteria isolated from the mesotrophic lake. Most of them showed a maximum activity after 24 hours of growth. In strains isolated from the eutrophic lake, such activity was noted after 48 or 144 hours of cultivation.

Endo- and exopolygalacturonases activity is shown in Table 3. Bacteria isolated from the mesotrophic lake had a higher endopolygalacturonase activity, but the exopolygalacturonase activity was higher in strains derived from the eutrophic lake. This would indicate a more active decomposition of pectin by organisms inhabiting eutrophic lakes than by those of the mesotrophic ones.

It appears from the data on the activity of the

185

enzymes studied that it was different in particular strains and depended upon the effect of temperature and pH applied in the course of the research. Figure 1 indicates the effect of temperature on proteolytic enzymes activity in strains of the genera *Pseudomonas, Achromobacter, Flavobacterium, Bacillus* and the family Enterobacteriaceae studied at pH 7.0. It may be noted that the highest activity of these enzymes in all strains appeared at 37°C. The maximum activity was noted in strains of the genus *Bacillus*.

Activity of amylolytic enzymes produced by the same bacteria studied at pH 7.0 and at different temperature is illustrated in Fig. 2. This activity was highest in strains of Pseudomonas and Bacillus at 37°C. In *Achromobacter* strains the above activity was highest at 26°C and in those belonging to the genus *Flavobacterium* at 30°C.

Figure 3 shows the exopolygalacturonase activity studied at pH 7.0 and different temperatures. It may be noted that the *Bacillus* and *Pseudomonas* strains showed the highest activity at 37°C and the remaining ones at 30°C. The results obtained in these studies indicate differences in the activity of exoenzymes synthesized by water bacteria. This activity depends upon the trophy of the lake as well as the place and time of isolation of the bacteria. pH and temperature exert a considerable effect on this activity. Lack of data in the literature on the activity of the exoenzymes studied in this work limits a fuller discussion of the results.

However, the results presented here provide information on the bacteria that produce exoenzymes for decomposing organic compounds and the factors which affect the above processes.

Acknowledgements

I express my best thanks to Prof. Dr. Edmund Strzelczyk for his help and critical appraisal of the manuscript.

References

Coler, E. A. & Gunner, H. B. 1969. The rhizosphere of an aquatic plant Lemna minor. Can. J. Microbiol. 8: 964–966.

Collins, V. A. 1960. The distribution and ecology of Gram-negative organisms other than Enterobacteriaceae in lakes. J. Appl. Bact. 3: 510–514.

Henrici, A. T. 1939. The distribution of bacteria in lakes. In: Problems of lake biology. Amer. Assoc. Adv. Sci. 10: 39–64.

Kuznetsov, S. I. 1951. Rol mikroorganizmov v obrazovanii sapropylnykh otłoženii. Mikrobiologya (USSR). 3: 245–255.

Nalewajko, C. & Lean, D. R. S. 1972. Growth and excretion in planktonic algae and bacteria. J. Phycol. 8: 361.

Nelson, N. 1944. A photometric adaptation of the Samogyi method for the determination of glucose. J. Biol. Chem. 153: 375–380.

Schacterle, G. R. & Pollack, R. L. 1973. A simplified method for the quantitative assay of small amounts of protein in biological material. Anal. Biochem. 51: 645–656.

Vallentyne, J. F. 1957. The molecular nature of organic matter in lakes and oceans with reference to sewage and terrestrial soil. Fisheries Board of Canada. 14: 33–82.

LAKE MOSSØ, DENMARK

BACKGROUND DATA

Latitude: 56°02'
Longitude: 2°48'
Altitude: 22.3 m
l (km) 10
b (km) 2
L (km)
L_D
Name of the main tributary River Gudena
Average inflow m³sec. 4.4
Average outflow m³/sec.
Theoretical retention time 2 years

origin:
z (m) 22
$\bar{z}$ (m) 9
$\sigma z/z$ 0.41
V (km³) 0.151
A (km²) 17
A' (km²)
$A' : A$

Geological characteristics: moraine deposits

Climatic conditions: coast-temperate

Average monthly temperature
Average precipitation/year 600 mm
Average sunshine duration
Main wind direction(s) westerly

Ice cover (days) 0–120
Average radiation/year
% of calm days

Cultural geography and demography:

Land usage of catchment area (%)

Usage of lake water including
recreation activities

Industrial   2
Agricultural 80
Meadows   4
Forest   10
unused   4
number of residents 25.000

inhabitants/km² 36

	min	max
Water temperature: (°C)	min 0	max 25
Secchi depths: (m)	min 0.70	max 3.55
Euphotic zone:	min	max
O_2 concentration	min	max

pH 8–9   Conductivity (μS) 290–370   Alkalinity (mval) 2–2.5
Average P-conc. 100 μg/l   Average N-conc. 1500 μg/l

Conditions of sediment

Gyttja

Dom. phytoplankton species:
Asterionella, Melosira, Stephanodiscus, Microcystis and Anabaena

Dom. zooplankton species:

Dom. macrophytes:
Phragmites, Scirpus, Typha

Dom. benthic organisms:

Fishes:

LAKE VILA, GIRONA, SPAIN

BACKGROUND DATA

Latitude: 42°07′10″N

Longitude: 2°44′50″E

Altitude: 175 m

l (km) 0.170

b (km) 0.081

L (km) 0.470

L_D 1.19

Name of the main tributary –

Average inflow m³/sec.

Average outflow m³/sec.

Theoretical retention time

origin: Karstic,
gypsiferous

z (m) 33

$\bar{z}$ (m) 4.64

$\bar{z}/z$ 0.58

V (km³) 57.764×10^{-6}

A (km²) 12.450×10^{-3}

A' (km²) 12.450×10^{-3}

$A' : A$ 1

Geological characteristics

Karstic. Dissolution of gypsum strata. Meromictic.

Climatic conditions

Average monthly temperature 15°C

Average precipitation/year 115 mm

Average sunshine duration 200 h/month

Main wind direction(s) N–S

Ice cover (days) 0

Average radiation/year 19120 cal cm^{-2}

% of calm days

Cultural geography and demography:

Land usage of catchment area (%)
  No
Industrial
Agricultural
Meadows
Forest
unused
number of residents

Usage of lake water including
recreation activities

inhabitants/km²

Water temperature: min 4°C   max 23°C
Secchi depths: min 2 m   max 5 m
Euphotic zone: min 6 m   max 6 m
O_2 concentration: min 0 ppm   max 8 ppm
pH 7.5   Conductivity (μS) 1100–2100   Alkalinity (mval) 11.5
Average P-conc. 2.5 μmols/l   Average N-conc. 20 μmols/l

Conditions of sediment:

Completely anaerobic

Dom. phytoplankton species:
Chlorella ellipsoidea
Dom. bacterioplankton species: Chromatium minus, Chlorobium phaeobacteroides in chemocline

Dom. zooplankton species:
Filinia longiseta, Tropocyclops pras-nus. Some Chaoborus sp.

Dom. macrophytes:
Phragmites communis, Sparganium sp.

Dom. benthic organisms:
No

Fishes:
Cyprinus carpio

LAKE SISO, GIRONA, SPAIN

BACKGROUND DATA

Latitude: 42°07′35″N

Longitude: 2°45′05″E
Altitude: 175 m
l (km) 0.026
b (km) 0.020
L (km) 0.077
L_D 1.03
Name of the main tributary –
Average inflow m³/sec.?
Average outflow 1/sec. 0.5
Theoretical retention time 0.22 years

origin: Karstic,
      gypsiferous
z (m) 11
$\bar{z}$ (m) 8.40
$\bar{z}/z$ 0.76
V (km³) 3.843×10^{-6}
A (km²) 0.457×10^{-3}
A' (km²) 0.457×10^{-3}
$A' : A$ 1

Geological characteristics:
Karstic. Dissolution of gypsum strata

Climatic conditions

Average monthly temperature 15°C
Average precipitation/year 115 mm
Average sunshine duration 200 h/month
Main wind direction(s) N–S
Evaporation per year 1080 m³

Ice cover (days) 0
Average radiation/year 19120 cal cm^{-2}
% of calm days always

Cultural geography and demography:

Land usage of catchment area (%)
  No catchment area
Industrial No
Agricultural
Meadows
Forest
unused
number of residents

Usage of lake water including
  recreation activities

inhabitants/km²

Water temperature: min 4°C   max 23°C
Secchi depths: min 0.35 m   max 2.30 m
Euphotic zone: min 3 m   max 5 m
O_2 concentration: min 0 ppm   max 5 ppm
pH 7.5   Conductivity (μS) 2160   Alkalinity (mval) 7.60
Average P-conc. 2.5 μmols/l   Average N-conc. 0.25 μmols/l

Conditions of sediment:
   Completely anaerobic

Dom. phytoplankton species:
   Euglena spp.
   Com. bacterioplankton species: Chromatium minus, Chlorobium phaeobacteroides, Chlorobium phyaeovibroides

Dom. zooplankton species:
   Some Chaoborus sp.
Dom. macrophytes:
   No
Dom. benthic organisms:
   No
Fishes:
   No

4.4 Zoological Aspects

COMPOSITION AND VARIATION OF THE BOTTOM FAUNA IN THE SUBLITTORAL OF THE EUTROPHIC LAKE DOIRAN (MACEDONIA, YUGOSLAVIA)

J. A. ŠAPKAREV

Abstract

The qualitative and quantitative composition of the bottom animals was investigated in the shell zone of the eutrophic lake Doiran. Also an attempt was made to show the seasonal changes in the population densities of the dominant species, their groups and all the benthal fauna during a one-year period. Finally, the composition of the sublittoral animal community and the population density of the dominant species in the different transects of the lake was researched.

Introduction

The shell zone is one of the most interesting phenomena of the lake sediment. The deposition of shells of the lake's dead molluscs is restricted to a very distinct bathymetric position, forming a continuous belt parallel to the lake's shore. The depositions of the shell zone represents a typical thanatocenosis in the sense of Wasmund (1926). The formation of the shell zone is due to the combined action of biotic and mechanical factors (Lundbeck, 1929).

In Europe this zone is observed in the great Balkan lakes as well as the Baltic ones (Wasmund, 1926; Lundbeck, 1929; Stanković, 1933) where it covers the sublittoral bottom. A typical representative of the eutrophic Balkan lakes is Lake Doiran in which the deposition of the molusc shells occupies the lake bottom between the isobaths of 4 and 6 meters. In this lake the shell zone is mainly composed of *Dreissena*

polymorpha and to a lesser extent of *Valvata piscinalis*. The shell zone varies in width and according to our observations in the Yugoslav part of the lake it becomes wider from northwest to southeast (see Fig. 1).

The shell zone of lake Doiran is populated with a particular animal community which will be described in this paper and which differs from the communities of the littoral and "profundal" zones (Šapkarev, 1975a; 1975b). Also presented are seasonal changes in the population densities of the dominant species and their groups as well as the dynamics of the densities of all the benthal fauna during a one-year period. Finally, investigations on the composition of the sublittoral community and the population density of the dominant animal species in different transects of the lake are included.

Method of investigations

From January through December 1964, quantitative samples were taken with an Ekman grab (15 × 15 cm) with a 6 meters long pole and an Ekman-Birge dredge modified by Borutzky having a relatively high weight. Samples were made at monthly intervals on a transect between Star and Nov Doiran. In July samples were obtained in five separate transects along the entire Yugoslav part of the lake. All benthic collections were taken from three isobathes ranging from 4 to 6 meters depth.

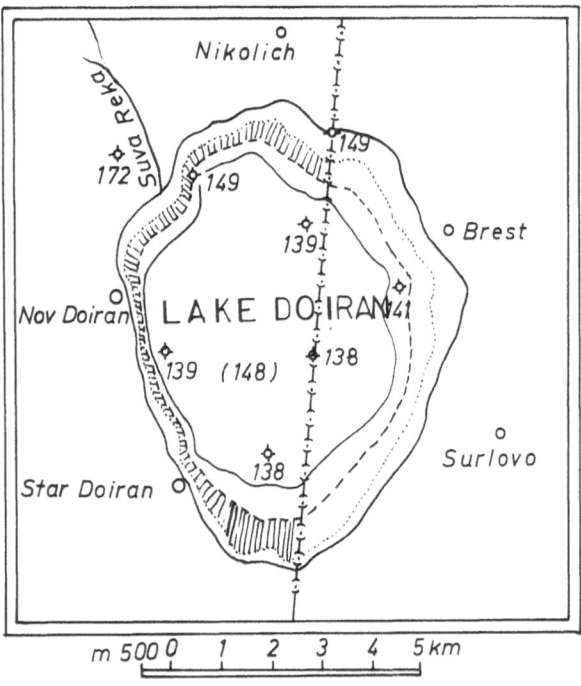

Fig. 1. Survey map of Lake Doiran (cross-hatching denotes the explored area of the lake).

Results of investigations

The composition of the benthic community of the sublittoral zone of Lake Doiran involves a great number of animal groups (see Table 1). This list is obviously incomplete because it contains animal groups where the taxonomical status of their species has not yet been determined, such as the groups Hydridae, Nematoda, Ostracoda, Hydracarina, Odonata and Trichoptera.

The dominant animal groups in terms of number of species are the chironomids (12 species), oligochaetes (11), and leeches (5), which together with the triclads and the crustaceans form the most significant part of the sublittoral animal community.

The dominant species of this community are *Dugesia polychroa* and *Eudendrocoelum lacteum* from the triclads, *Dreissena polymorpha* from the mussels, *Potamothrix hammoniensis* and *Criodrilus lacuum* from the oligochaetes, *Erpobdella octoculata, Glossiphonia complanata* and *Hemiclepsis marginata* from the leeches, *Asellus aquaticus* and *Rivulogammarus triacanthus* from crustaceans, *Procladius* sp. and *Chironomus* gr.

plumosus from the chironomids, *Chaoborus crystallinus* from the culicids and *Palpomia* sp. from the ceratopogonids.

During 1964 *Dreissena polymorpha* exhibited the greatest population density (355.2–5061.6 ind m^{-2}. The densities of other dominant species are as follows: crustaceans *Asellus aquaticus* (177.6–1465.2) and *Rivulogammarus triacanthus* (177.6–843.6), oligochaetes *Potamothrix hammoniensis* (222.0–932.4) and *Criodrilus lacuum* (44.4–754.8), the chironomid *Chironomus* gr. *plumosus* (44.4–666.0), the culicid *Chaoborus crystallinus* (44.4–355.2) and the triclads *Dugesia polychroa* (44.4–310.8) and *Eudendrocoelum lacteum* (44.4–266.4).

Dreissena polymorpha is very numerous in Lake Doiran (Šapkarev, 1969) having one population peak in the littoral zone, composed mainly of young individuals, with a second peak in population numbers in the sublittoral zone composed almost exclusively of adult specimens attached to empty shells or each other, often forming clusters of individuals.

The two mentioned crustaceans also show two peaks of population density, one occurring in the littoral and the other in the sublittoral (Šapkarev, 1975; Šapkarev & Angelovski, 1977). The same situation occurs with the triclads. The oligochaete *Potamothrix hammoniensis* maintains its maximum density in the deepest waters of the lake while *Criodrilus lacuum* shows two maxima of density – one in the shallowest water and one in the shell zone. The leech *Erpobdella octoculata* shows its maximum density in the littoral. The larvae of *Chironomus* gr. *plumosus* and *Chaoborus crystallinus* show their maximum populations in the deepest waters.

The deposits of shell which characterize the sublittoral bottom serve as attachment sites for the sponges, the mollusk *Dreissena polymorpha* and the leeches *Erpobdella octoculata, Glossiphonia complanata, Hemiclepsis marginata* and *Helobdella stagnalis* and the triclad egg sacks. In addition these shells serve as shelter areas for creeping triclads, oligochaetes, crustaceans and the other animals (listed in Table 1).

Considering its composition, the animal community of the sublittoral zone as to have an intermediary character. A good number of animal

196

Table 1. Average values of the population density (individuals per square meter) of different animal species in the composition of the sublittoral benthal community in different transects of Lake Doiran.

species / transects	between Star & Nov Doiran	Star Doiran - NATO	Skelet - Brest	Suva Reka - Surlovo	Karaula - Nikolić
SPONGILLIDAE					
Ephydatia fluviatilis L.	-	-	-	-	+
RHABDOCOELA indet.	-	-	-	-	22,2
TRICLADIDA					
Eudendrocoelum lacteum Müll.	-	333,0	-	-	44,4
Dugesia polychroa O. Schmidt	-	466,2	-	-	88,8
HYDRIDAE indet.	-	-	-	-	22,2
NEMATODA indet.	-	-	-	44,4	-
VALVATIDAE					
Valvata piscinalis Müll.	-	66,6	44,4	-	22,2
DREISSENIIDAE					
Dreissena polymorpha Pall.	2264,4	2020,2	1136,6	-	1043,4
NAIDIDAE					
Nais barbata Müll.	22,2	-	-	-	-
TUBIFICIDAE					
Potamothrix hammoniensis Mich.	777,0	532,8	732,8	1953,6	1354,2
Isochaeta dojranensis Hr.	22,2	-	-	-	-
Psammoryctes oligosetosus Hr.	44,4	66,6	222,0	-	177,6
Psammoryctes albicola Mich.	-	-	-	66,6	66,6
Psammoryctes moravicus Hr.	110,0	-	-	88,8	-
Peloscolex velutinus Grube	-	532,8	-	-	-
Limnodrilus hoffmeisteri Clap.	-	44,4	-	88,8	-
Limnodrilus udekemianus Clap.	-	22,2	-	66,6	-
Aulodrilus pigueti Kow.	111,0	-	-	-	-
GLOSSOSCOLECIDAE					
Criodrilus lacuum Hoffm.	-	88,8	199,9	88,8	-
GLOSSIPHONIIDAE					
Glossiphonia complanata L.	-	44,4	-	-	199,8
Glossiphonia heterocliata L.	-	111,0	-	-	-
Hemiclepsis marginata O.F. Müller	-	-	-	-	44,4
Helobdella stagnalis L.	-	44,4	44,4	-	44,4
ERPOBDELLIDAE					
Erpobdella octoculata L.	599,0	177,6	266,4	22,2	1354,2
ASELLIDAE					
Asellus aquaticus L.	999,0	1065,6	577,2	-	2597,3
GAMMARIDAE					
Rivulogammarus triacanthus Sch.	421,8	1176,6	954,6	44,4	532,8
ATYIDAE					
Atyaephyra desmarestii Millet	-	22,2	-	-	-
OSTRACODA indet.	66,6	88,8	-	-	-
HYDRACARINA indet.	199,8	-	..	-	-
CANEIDAE					
Caneis macrura Fabr.	22,2	-	-	-	66,6
ODONATA indet.	-	22,2	-	-	-
TRICHOPTERA indet.	-	22,2	-	-	-
CHIRONOMIDAE					
Tanytarsus gr. gregarius Kieff.	22,2	-	-	-	-
Procladius sp.	22,2	-	88,8	377,4	-
Pelopia kraatzi Kieff.	22,2	88,8	-	-	66,6
Chironomus gr. plumosus L.	22,2	-	-	-	-
Limnochironomus gr. nervosus Staeg.	-	-	-	-	22,2
Cryptochironomus gr. defectus Kieff.	22,2	-	133,2	-	-
Cryptochironomus gr. viridulus F.	-	-	-	44,4	-
Cryptochironomus gr. conjugens Kieff.	44,4	44,4	133,2	-	22,2
Polypedulum gr. nubeculosum Mg.	-	44,4	-	44,4	-
Polypedulum gr. convictum Walk.	22,2	-	-	-	-
Crycotopus gr. silvestris F.	-	-	-	44,4	-
Psectrocladius gr. psilopteras K.	-	-	-	-	22,2
CULICIDAE					
Chaoborus crystallinus de Geer	88,8	-	88,8	244,2	-
HELEIDAE					
Palpomya sp.	-	66,6	68,3	466,2	66,6

species are found there which normally either live predominantly in the littoral (sponges, *Dugesia polychroa, Eudendrocoelum lacteum, Isochaeta dojranensis, Nais barbata, Limnodrilus hoffmeisteri, L. udekemianus, Glossiphonia complanata, Helobdella stagnalis, Polypedulum* gr. *nubeculosum, Pelopia craatzi, Psectrocladius* gr. *psilopterus, Limnochironomus* gr. *nervosus* and *Dreissena polymorpha*) or in the "profundal" (*Potamothrix hammoniensis, Chironomus* gr. *plumosus, Tanytarsus* gr. *gregarius* and *Chaoborus crystallinus*). In contrast to Ohrid Lake, the sublittoral community of Lake Doiran is not characterized by animal species limited in their vertical distribution to the sublittoral zone nor do they attain the maximum density of their populations there.

The seasonal changes of Lake Doiran sublittoral benthic fauna are evident. They are illustrated by the dominant representatives as well as the different animal groups and the total benthic fauna.

The diagrams in Fig. 2 illustrate the quantitative variations of the two triclads, *Dugesia polychroa* and *Eudendrocoelum lacteum*, in the course of the year. The maximum density of both of them was reached in June–July, the minimum – in winter (December – January). The same variation was observed during 1967 from the stony habitat of the littoral zone in this lake (Šapkarev, 1975b). This can be explained by the seasonal character of reproduction which, with both species occurs in the spring months, so that a great number of young individuals are seen in June–July, i.e. at the end of spring and beginning of summer (Šapkarev, 1972).

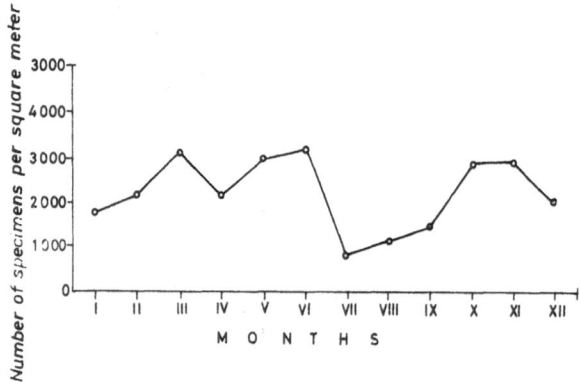

Fig. 3. Dynamics of the population density of *Dreissena polymorpha* Pall. during 1964 from the sublittoral zone of Lake Doiran.

The rhythmic seasonal changes of *Dreissena polymorpha* are shown in Fig. 3. The diagram shows the minimum population density during the summer (July–September). This minimum could be explained first of all by the appearance of anaerobic conditions which may last several days or even longer, resulting in the death of a large percentage of *D. polymorpha*. During 1972 in the littoral stony habitat the minimum population density of this mollusk was found in May–June (Šapkarev, 1975b).

The curves of Fig. 4 show the seasonal variation of the population density of two oligochaetes, *Potamothrix hammoniensis* and *Criodrilus lacuum*. The first species exhibits during the year three maxima while the second shows only one, in June, as a result of the appearance of a new generation. In the deepest waters of this lake *P. hammoniensis*

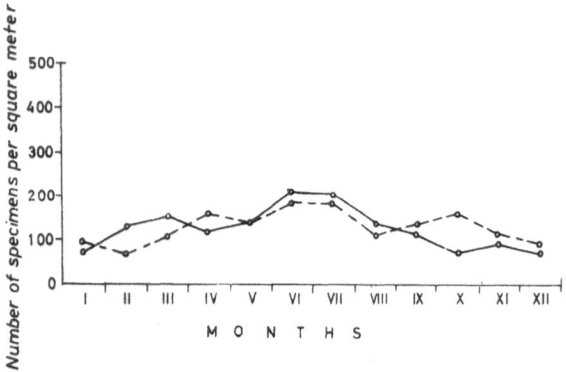

Fig. 2. Dynamics of the population density of two triclads during 1964 from the sublittoral zone of Lake Doiran.
——— *Dugesia polychroa* – – – – – – *Eudendrocoelum lacteum*

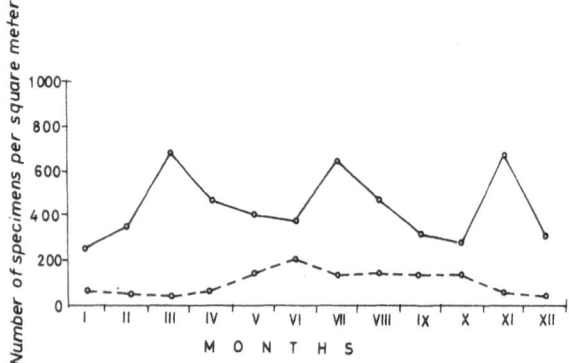

Fig. 4. Dynamics of the population density of two oligochaetes during 1964 from the sublittoral zone of Lake Doiran.
——— *Potamothrix hammoniensis* – – – – – – *Criodrilus lacuum*

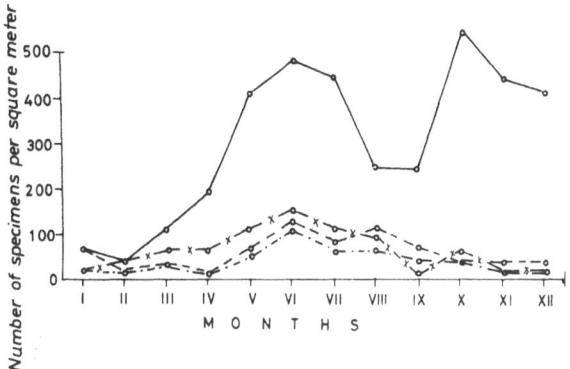

Fig. 5. Dynamics of the population density of four leeches during 1964 from the sublittoral zone of Lake Doiran. ——— *Erpobdella octoculata* - - - - - - *Glossiphonia complanata* –.–.–.– *Helobdella stagnalis* — × — × — *Hemiclepsis marginata*

has shown two maxima during the year (Šapkarev. 1975a).

Figure 5 shows the variations of the population density of four leeches. All of them show a maximum density in June (except *Erpobdella octoculata* which exhibited two maxima during the investigation period). In the stony habitat of the littoral zone these leeches have shown during 1968/69 a maximum density in July and a minimum in February (Šapkarev, 1975b). According to previous investigations (Šapkarev, 1970; 1971) this maximum is a result of the reproductive period which mainly occurs in spring with newly hatched leeches appearing from May to June.

The dynamics of the population densities of two crustaceans, *Asellus aquaticus* and *Rivulogam-*

marus triacanthus, are represented in Fig. 6. Their maximal densities appeared in May, October and November respectively. The variation of the population density of *A. aquaticus* in the shell zone of this lake during 1976 also shows two maxima but in March and December and a minimum density from June to September (Šapkarev & Angelovski, 1977).

Figure 7 shows the seasonal variations of the density of populations of two chironomids, *Chironomus* gr. *plumosus* and *Procladius* sp., and one culicid *Chaoborus crystallinus*. It should be noted that its inverse relationship has been observed during the investigated year between *Ch.* gr. *plumosus* and *Procladius* sp. The minimum population density of *Ch.* gr. *plumosus* was found in May–June when *Procladius* sp. was at its maximum population size. During five-year investigation in the deepest waters of Lake Doiran, the minimal population density of *Ch.* gr. *plumosus* has appeared in July or August while the minimal density in September or October (Šapkarev, 1975a). According to Šapkarev (1968) the minimal population density of *Ch.* gr. *plumosus* is due to the eclosion while the maximal density to the appearance of the new generation. *Ch. crystallinus* shows minimum population density during the spring months and maximum during the end of autumn and the beginning of winter. Almost the same situation was found in the deepest waters of the lake (Šapkarev, 1975a).

The rhythmic seasonal changes of Lake Doiran's sublittoral benthic animal groups and the total

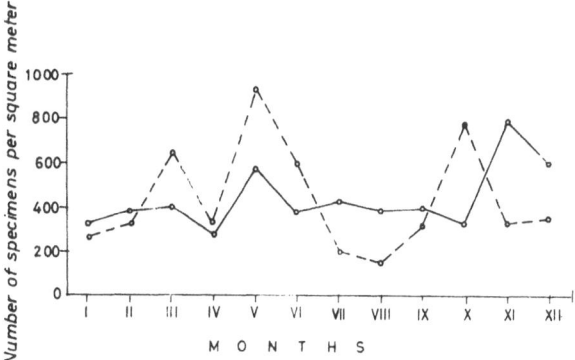

Fig. 6. Dynamics of the population density of two crustaceans during 1964 from the sublittoral zone of Lake Doiran. ———*Rivulogammarus triacanthus* - - - - - - *Asellus aquaticus*

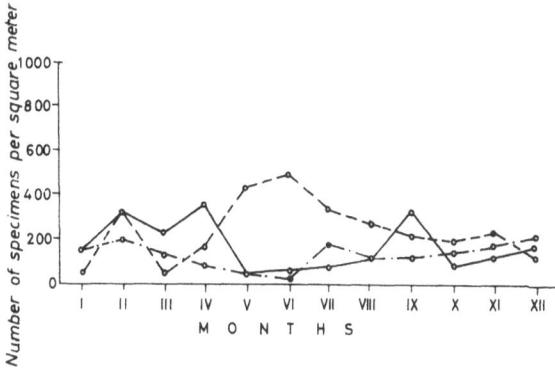

Fig. 7. Dynamics of the population density of two chironomids and one culicid during 1964 from the sublittoral zone of Lake Doiran. ——— *Chironomus plumosus* - - - - - *Procladius* sp. –.–.–.– *Chaoborus crystallinus*

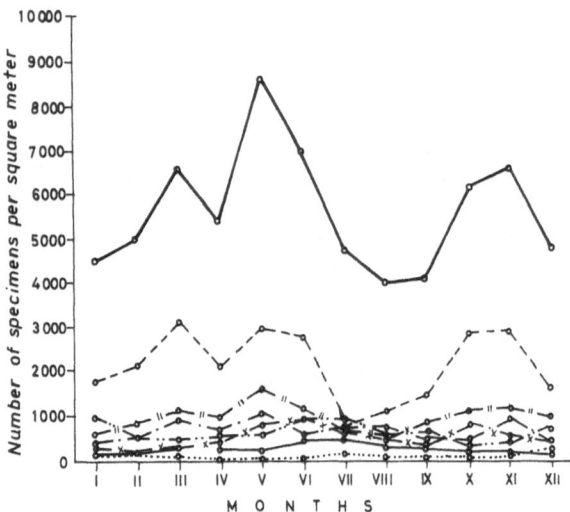

Fig. 8. Dynamics in number of specimens of the entire sublittoral benthal fauna and its animal groups of Lake Doiran during 1964. ————Tricladia - - - - - - Bivalvia -.-.-.- Oligochaeta —×—×—×—×— Hirudinea —"—"—"—" Crustacea —··—··— Chironomidae ————macrozoobenthos

benthic animal community are represented in Fig. 8. It is obvious that Bivalvia (more than 99% belonging to *Dreissena polymorpha*) predominates quantitatively in the benthic community of Lake Doiran sublittoral. This group does not go below an average of 2200 individuals per square meter, which is 38.8% of the total number of individuals of the entire community of this zone. Together with the next dominant group, Crustacea (about 1000 individuals per square meter in average), they comprise 56% of the sublittoral benthic community.

The relative densities of populations of other benthic groups during the investigated year is illustrated in Fig. 8. The groups such as Gastropoda, Hydracarina, Nematoda, Odonata and Heleidae are not represented in Fig. 8 because they were rarely observed in the sublittoral benthic community.

The seasonal changes of the entire sublittoral benthic fauna of Lake Doiran show two maximal densities – one in spring (March–June) and another in autumn (October–November) and two minimal densities – one in summer (July–September) and another in winter (December–January). This is determined by the seasonal variations of densities of some animal groups, especially the predominant ones such as Bivalvia and Crustacea, which are illustrated in Fig. 8.

The benthic fauna found in the different transects of the lake's sublittoral are qualitatively and quantitatively shown in Table 1. From this table it is obvious that the density of the total benthic animal community is different in the different transects. The density is greatest in the transect Karaula – Nikolić (in average 7925.1 individuals per square meter or maximum of 12032.4 individuals per square meter at the depth of 6 meters) and the smallest in the transect Suva Reka-Surlovo (in average 3640.8 individuals per square meter or minimum of 3285.6 individuals per square meter at the depth of 5 meters). In our opinion it is conditioned by the degree of development of the shell zone to the transect. The density of the animal settlement is bigger there whereas the shell zone is better developed, i.e. wider. The dominant groups Dreisseniidae, Oligochaeta, Hirudinea, Crustacea and Chironomidae are more abundant in the transects between S. & N. Doiran, S. Doiran-NATO, Karaula-Nikolic and Skelet-Brest, i.e. in the south-west part of the Yugoslav territory of the lake than the transect Suva Reka-Surlovo, i.e. north part of the lake.

The dominant species *Dreissena polymorpha*, *Erpobdella octoculata*, *Asellus aquaticus* and *Rivulogammarus triacanthus* densely populate all transects except Suva Reka-Surlovo where D. polymorpha is absent in the sublittoral. All transects are densely populated with *Potamothrix hammoniensis* however.

On the basis of the data quantitatively collected from the different transects of the lake, the sublittoral zone of Lake Doiran for the period investigated was populated with an average of 5857.5 individuals per square meter or on the basis of the data from the transect searched during the one-year period for 1964 was 5615.3 individuals per square meter.

References

Lundbeck, J. 1929. Die "Schalenzone" der norddeutschen Seen. Jhrb. d. Preuss. Geol. Landesanst., 49: 1127–1151.

Šapkarev, J. 1968. Ecology and dynamics of the population and biomass of *Chironomus plumosus* (Diptera: Chironomidae) in Lake Dojran, Macedonia. Izdanija, Inst. piscic. RSM, Skopje, 4:1–40 (in Yugoslav).

Šapkarev, J. 1969. Distribution and the population density of *Dreissena polymorpha* Pall. (Lamellibranchia) in the lakes of Macedonia. Ann. Fac. Sci. Univ., Skopje, 21:31–52 (in Yugoslav).

Šapkarev, J. 1970. Seasonal changes in a population of *Erpobdella octoculata* L. (Hirudinea) in the large lakes of Macedonia (Dojran, Prespa and Ohrid). Ann. Fac. Sci. Univ., Skopje, 22:19–31 (in Yugoslav).

Šapkarev, J. 1971. Idioecological investigations of some leeches in lakes Dojran, Prespa and Ohrid. Macedonia. Ekologija, Beograd, 6:105–112 (in Yugoslav).

Šapkarev, J. 1972. Seasonal changes in populations of *Dendrocoelum lacteum* Mull. and Dugesia polychroa O. Schmidt (Turbellaria: Tricladida) in the shore stone habitat of Lake Dojran. Ekologija, 7: 183–195 (in Yugoslav).

Šapkarev, J. 1975a. Seasonal and annual variation of the population density and biomass of the bottom fauna in the deepest waters of Lake Dojran, Macedonia. Sym. Biol. Hung., Budapest, 15: 255–263.

Šapkarev, J. 1975b. Composition and dynamics of the bottom animals in the littoral zone of Dojran Lake, Macedonia. Verh. Internat. Verein. Limnol., Stuttgart, 19: 1339–1350.

Šapkarev, J. & Angelovski, P. 1977. The ecology and distribution of *Asellus aquaticus* L. (Isopoda: Asellidae) in Lake Dojran. Ann. Fac. Biol., Skopje, 30: 57–77.

Stanković, S. 1933. La zone a coquilles des lacs balkaniques du sud. Rec. trav. offeret J. Georgievitch, Beograd, 233–253 (in Yugoslav).

Wasmund, E. 1926. Biocönose und Thanatocönose. Arch. Hydrobiol., 17: 1–116.

GENERAL INDEX

206

equitability, 74, 75, 76
erosion, 43, 56
   water, 7 ff.
   wind, 43, 56
Eudendrocoelum lacteum Müll., 196 ff.
Euglenophyta, 124
euphotic zone, 11, 16, 65, 70
eutrophication, 11, 13, 35, 41, 55, 73, 77, 83, 87, 112, 128, 136
eutrophic, 3, 11, 16, 63, 81, 84, 97, 99, 105, 114, 121, 122, 123, 125, 127, 131, 132, 137, 138, 153, 183, 184, 185, 195
evaporation, 89
excretions, 183

fall overturn, 169
feeding (f.-habits), 11, 49, 84, 85, 133, 136, 138
   conditions, 81
   rate, 81
filtration/filtering, 55, 57, 98, 104, 105, 156, 170, 174
fish, 82
   feeding, 11
   fry, 49
   planktivorous, 46, 49, 50, 57
   production, 81
flagellates, 121
Flavobacterium sp., 184, 186
flux-carbon, 153
   quantum, 20
food, 45, 83 ff., 131, 183
   chain, 50
   composition, 81, 86
   intake, 84, ff.
   resource, 81, 86
   supply, 86
   web, 67
Fragillaria, 101
   crotonensis, 104
fry, 49
fungi, 153

Gammaridae, 197
Gloeocystis planktonica, 124
Glossiphonia complanata, 196 ff.
   heterocliata, 197
Glossiphoniidae, 197
Glossoscolecidae, 197
Gomphosphaeria lacustris, 77
gradient light, 21, 43, 161, 173
   physico chemical, 170, 179
   sulfide, 180
grazing, 101, 126, 130, 136
green algae, see chlorophyta
gross-load, 55

potential, 57
growing season, 58
growth, 139
   algal, 99, 112
   bacteria, 161, 174
   benthos, 67
   limitation, 101, 144, 180
   phytoplankton, 145
   plankton, 67
   population, 174

hardness, 90, 91
Heleidae, 197, 200
Helobdella stagnalis L., 196 ff.
helophytes, 128, 132 ff.
   herbaceous, 131 ff. 136
   reedy, 132, 133, 136, 138
Hemiclepsis marginata O. F. Müller, 196, 197
heterotrophic bacteria, 169, 183, 185
Hirudinea, 200
holomictic, 161, 163, 169, 173
humic-lakes, 121, 128 ff.
   matter, 36, 90
Hydracarina, 196, 197, 200
Hydridae, 196, 197
Hydrogen sulfide, (H_2S), 12, 129, 164, 165
hydrophytes, 128, 132, 134
hypertrophic, 63, 77, 81
hypolimnion, 129, 163

ice (cover), 3, 65, 73, 111, 112, 115, 119, 122, 144, 145
ignition loss, 4, 123
inflow, 37, 55, 57, 97
   subsurface, 56, 57
   sulfate, 169
   superficial, 37
   surface, 41, 55, 57
   underground, 16, 89
input, 17, 40, 54, 58
   allochthonous, 11
   birds, 54, 55
   nutrient, 41, 53, 122, 132, 145
   phosphorus, 53, 54
   sewage, 99
   water, 98
interface water/sediment, 8, 54, 63, 72
interphase sediment/water, 162
ionic-composition, 89, 91, 94
   proportion, 91
Iris pseudacorus, 137
Iron (Fe), 3, 7, 8, 12, 53, 55, 56, 57, 90, 123, 140
   sulfate, 55
Isochaeta dojranensis Hr., 197, 198

207